Technology and
Social Change in Rural Areas

Other Titles in This Series

Also of Interest

† Available in hardcover and paperback.

Rural Studies Series

Technology and Social Change in Rural Areas
A Festschrift for Eugene A. Wilkening
edited by Gene F. Summers

The possibility of nuclear war, the failure of the Green Revolution, the capabilities of genetic engineering, and other actual and potential effects of technological innovations have created demands for a more humane application of technology. Addressing this issue, *Technology and Social Change in Rural Areas* is a clear assessment of the current state of affairs. The book begins with a discussion of the changing paradigms of technology adoption and diffusion, the dynamics of public resistance, and the question of social responsibility in an age of synthetic biology. In subsequent sections, the contributors assess the revolutionary effect of technology on agriculture worldwide and conclude that radically new public policies are essential; expose the transformations of rural life and communities that result from the localized effects of technology and its use as a weapon in world-system politics; and critically examine the appropriate technology movement.

The essays are presented to honor Professor Eugene A. Wilkening for his many pioneering and lasting contributions to the study of technology and rural social change. The book includes an intellectual biography of Professor Wilkening written by his long-time colleague and friend, William H. Sewell.

Gene F. Summers is professor of rural sociology, University of Wisconsin—Madison. His work on community changes associated with the restructuring of national economies has resulted in two books, *Industrial Invasion of Nonmetropolitan America* and *Nonmetropolitan Economic Growth and Community Change*.

Technology and Social Change in Rural Areas

edited by Gene F. Summers

A Festschrift for Eugene A. Wilkening

Routledge
Taylor & Francis Group
LONDON AND NEW YORK

First published 1983 by Westview Press

Published 2019 by Routledge
52 Vanderbilt Avenue, New York, NY 10017
2 Park Square, Milton Park, Abingdon, Oxon OX14 4RN

Routledge is an imprint of the Taylor & Francis Group, an informa business

Copyright © 1983 Taylor & Francis

Library of Congress Catalog Card Number 83-50118

ISBN 13: 978-0-367-28966-9 (hbk)

ISBN 13: 978-0-367-30512-3 (pbk)

Contents

Part 3 Technology and Rural Life

Part 4 The Appropriate Technology Movement

Acknowledgments

We wish to express our appreciation to David L. Brown and other members of the Rural Studies Series Editorial Board for their encouragement and helpful guidance in the preparation of the Festschrift. Financial support was provided by the Rural Sociological Society, the College of Agricultural and Life Sciences of the University of Wisconsin and the Department of Rural Sociology; these funds have made possible this celebration of Gene Wilkening's accomplishments.

For their many hours of skilled and cheerful efforts in the preparation of the manuscript, I want to express my personal appreciation to Caty Ahrens, Kathy Reinhard and Lorna Miller. A very special debt of gratitude is owed to Doris Slesinger for her excellent skill in proofreading.

G.F.S.

Eugene A. Wilkening:
A Biographical Note

Eugene A. Wilkening was born and raised on a farm in Southeastern Missouri. His father was a hard-working, business-like man who operated a 200-acre general farm. His mother, in addition to her household and farm duties, was a rural school teacher. Gene attended a nearby one-room school through eighth grade and went on to high school in the nearby Village of Oak Ridge from which he was graduated in 1933 in a class of 13 students. It was his mother who insisted that the children go on to higher education. All four graduated from college. The three boys eventually earned Ph.D.'s and have had highly successful academic careers. Gene first attended Cape Girardeau State Teachers College. At the end of his sophomore year he decided to transfer to the University of Missouri to study horticulture. To help pay his expenses he worked in the laboratories of the Veterinary and the Horticulture Departments. He fully intended to devote his life to horticulture and was particularly interested in the research that was being done on new crop varieties. In fact his first experience with the diffusion of new farm practices came when he brought home some of the new hybrid seed corn from the University and induced his reluctant father to plant it. His father became an early champion of hybrid corn once the harvest was in and his yields had increased.

Gene, however, was not destined to become a horticulturalist. After taking courses in agricultural economics and rural sociology, he shifted his major to agricultural economics because, as he said later, he had become "more interested in people than in things." He was awarded a Bachelor of Science degree in 1937. By that time the various "New Deal" agencies, which had been set up to cope with the seriously depressed condition of agriculture, were functioning throughout rural America. Foremost among these was the Rural Rehabilitation Program of the Farm Security Administration. Gene took a position in this agency helping economically distressed farmers obtain financial assistance through its "substandard" loan program in Missouri.

During this period he also began graduate work in rural sociology at Missouri. To gain experience and earn money for his expenses, he corrected papers for correspondence courses in rural sociology. His major professor was Charles Lively, who was one of the leading rural sociologists of his generation and was one of the founding fathers and early Presidents of the Rural Sociological Society. Gene's Master's thesis on geographic mobility of Farm Security clients grew out of one of Lively's research projects.

Gene then took a position as an interviewer on a study sponsored by the Bureau of Agricultural Economics of the USDA in cooperation with the Department of Rural Sociology under Lively. This was a study of share-croppers in Southeastern Missouri who had been displaced by their owners and had no alternative other than to become farm laborers. Looking back over this experience many years later, he says that he learned a great deal from Lively about the definition of research problems, the design of questionnaires, interviewing, quantitative analysis, and especially about the importance of careful report writing.

During this period Gene also had an opportunity to work with Cecil Gregory on a project that involved planning for the relocation of farm families who were to be displaced by the construction of a flood control dam on the Saint Francis River. Again, he did much of the interviewing and analysis on this study and co-authored (with Gregory) his first publication, *Planning For Family Relocation in the Wappapello Dam Basin Area* (Missouri Agricultural Experiment Station Bulletin No. 427, 1941).

Before the report was published, the United States had entered World War II and Gene, like many others of his generation, was faced with the draft and with serious problems of conscience, which he temporarily resolved by becoming a conscientious objector. In the summer of 1941 he entered the University of Chicago to begin graduate study, where he took courses with Louis Wirth, Robert Redfield, Herbert L. Blumer, Everett Hughes, William F. Ogburn, Earnest Burgess, and Lloyd Warner. Toward the close of his first year at Chicago, he learned from Wirth about the treatment of Jews under Hitler and after much thought decided to write his draft board to change his conscientious objector classification. Within a few days he was ordered to report to Jefferson Barracks for induction into the Army where he remained to interview and assign new recruits. After one year he was sent to the Adjutant General School for officer training, graduating as a Second Lieutenant. He was then assigned as a personnel officer with the Air Corps assigned to a Troop Carrier Group and sent to England where he spent 18 months. Then, after brief service in France, he became a member of an advanced party in Berlin where he helped in the preparations for the Potsdam Conference. Following his release from service in 1946 he was hired by C. Horace Hamilton at North Carolina State College as an Instructor in Rural Sociology.

In the Fall of 1947 Gene returned to Chicago to work on his Ph.D. in Sociology and was awarded a Social Science Research Council Fellowship. That year he met and married Helena "Bunny" Emerson (the daughter of the noted Taxonomist Alford Emerson). Among his contemporaries in graduate school were several who also later became illustrious sociologists including Otis Dudley Duncan, Albert J. Reiss Jr., Morris Janowitz, Eleanor Bernert Sheldon, Guy E. Swanson and Neal Gross. It was Gross who had the greatest influence on Gene. Gross had come to Chicago from Iowa State College where he had worked with Bryce Ryan on their pioneering study of the diffusion of hybrid seed corn among Iowa farmers. Gene's latent interest in the adoption of farm practices was aroused by frequent conversations with Gross about the Iowa research. Gene decided to do a dissertation on this topic but encountered some difficulty in finding someone in that noted center of urban research who was willing to serve as his dissertation advisor. Herbert Blumer, whose course in social psychology had stimulated Gene, did not feel that he knew enough about the topic or about rural life to supervise a dissertation on the adoption of farm practices. Louis Wirth, who had already written his classic article on the urbanization of rural life and who was a man with broad sociological interests, consented to be Gene's advisor. The political scientist Charles Hardin, who had done extensive research on the Tennessee Valley Authority programs also served on the dissertation committee and provided valuable assistance. By the end of the academic year, Gene had completed his course work, passed his preliminary examinations and had his dissertation proposal approved. He then went back to North Carolina to do the field work on acceptance of farm programs and practices in a Piedmont community. He successfully defended the dissertation and was awarded the Ph.D. in 1949. This led to a larger study of farm practice adoption published under the title, *Acceptance of Improved Farm Practices in Three Coastal Plain Counties* (North Carolina Agricultural Experiment Station Technical Bulletin No. 98, 1952).

I first met Gene through Neal Gross who had earlier told me about the North Carolina study. I was then chairman of the Rural Sociology Department at Wisconsin and was anxious to change its directions and to recruit some strong younger scholars to replace professors who were nearing retirement — particularly our most eminent scholar, John H. Kolb. With the unplanned extension of a foreign leave granted to George Hill, I was authorized to bring Gene to Wisconsin as a temporary replacement. Toward the end of the year the Dean agreed to expand the department by one member, and we offered Gene the post as an Assistant Professor. He accepted and has been at Wisconsin ever since. He was promoted to Professor and became chairman of the department in 1955 and served through 1959, a period marked by great growth and development. Bringing Gene to Wisconsin was probably the most important single step ever taken to modernize the research and graduate teaching program of the department.

Gene's career at Wisconsin has been marked by prodigious productivity in a number of research areas. But the great bulk of his research is in the following areas: the adoption of improved farm practices, the roles of county extension agents, family farm relationships, and most recently the quality of life in the rural environment. In all of this work there is a consistent interest in technological innovations and social change with particular attention to the influence of the family and of social psychological factors on the behavior of rural people and agencies.

Probably the body of research for which Gene is most widely recognized is that on the adoption of farm practices. This is not surprising since he pioneered in this area and more than a third of his many publications have dealt with one or another aspect of this complex topic. It would be hard to single out a particular publication as his most important contribution in this area but to me the most impressive are those dealing with the role of the family in the decision process. This is a theme that runs through nearly all of his research on the adoption of new ideas and practices. Probably nothing that he or any other scholar has done has contributed more to make us understand that the adoption of farm practices is a family decision process; a process that is participated in by husband, wife, and children in varying degrees according to the values and goals of the family and the aspirations, skills, work roles, and responsibilities of the members, rather than on the basis of traditional notions about the division of authority in the farm family. This is not to say that he ignored the importance of outside communications and of communicating agents in bringing about adoption of practices and ideas — indeed several of his publications deal specifically with this topic — but rather that he correctly perceived that adoption of new ideas and technology is a family decision and sought to further understand the role of the family and its members in the decision process.

Gene's work on roles of County agents as viewed by the agents themselves, their clientele, and their supervisors was a pioneering research undertaking of great practical and theoretical importance. The work was begun in the early 1950's and published throughout the decade. It reflected the then current emphasis in social psychology and sociology on role theory and role conflict. Gene's research, reported in several monographs and articles, showed how selection, training, and personal characteristics of the agents influenced the agents' conception of his/her role, role performance, and role conflict and how discrepant role expectations on the part of supervisors, local leaders, and clients generated feelings of stress on the part of the agents. This study, and one conducted by Neal Gross and his associates on school superintendents (*Explorations in Role Analysis,* New York:Wiley & Sons, 1958) are among the most ambitious and enlightening field studies of social role and role conflict.

Gene's interest in the effect of planned interventions and change began while he was a graduate at Missouri working on the Wappapello dam project and has continued throughout his career. It is reflected in research monographs and articles on these topics dealing with Wisconsin and Brazilian communities. Perhaps this interest is best exemplified in his ongoing study of the Kickapoo dam flood control project in Wisconsin by which many farmers, villagers, and merchants have been or will be displaced when the dam is completed. He has followed the planning process, the environmental impact studies, the court battles, attended meetings of townspeople, farmers and governmental officials and has interviewed farmers and townspeople before and after they were relocated. Several publications have reported the results of his research. The dam has not yet been completed.

Gene's most recent interest in research on the quality of life began in the mid-sixties when he did a study in Brazil, supported by the Land Tenure Center, on the adjustments of migrants into central Brazil. This interest received greater impetus when in the early 1970's he conducted a study at the request of the Agricultural Extension Service on the effects of Rural Development in Northwestern Wisconsin. This was a project undertaken by the Extension Service to develop the industrial, agricultural, and recreational potential of 19 rural counties in Northwestern Wisconsin. Much of this area originally had been heavily forested and contained a scattering of lakes but had been cut over in the period of intensive lumbering in the Great Lakes states at the turn of the century. Soon after deforestation the thin lands had been cleared and settled by aspiring farmers but soon petered out leaving abandoned farmsteads and a great deal of rural poverty in the countryside and in the communities. With the second growth of timber and the development of paved highways, it was felt by community leaders and the Extension Service that the time was ripe for redevelopment of the area with major emphasis on recreation and limited development of farming and rural industry. Gene obtained a grant from the National Institute of Mental Health to study changes in the quality of life of the people of the area resulting from the redevelopment project. Following the lead of Angus Campbell and his associates at the Social Research Institute of the University of Michigan, he sought information on various subjective indicators of the quality of life of the people in the region. The anticipated and unanticipated consequences of the development program on the lives of the individuals and families of the area have been reported in a series of monographs and articles and need not be reviewed here because Gene has recently provided a comprehensive review of this research and the related work of others ("Subjective Indicators and the Quality of Life" in Robert M. Hauser, David Mechanic, Archibald Haller, and Taissa Hauser, eds. *Social Structure and Behavior,* New York: Academic Press, 1982).

This somewhat brief examination of Gene's major research efforts does not cover all of his contribution. He has also published articles and monographs on environmental attitudes, farmers' political participation, theory of family power, response stability in social surveys, distance and intergenerational ties of farm families, and large scale farming. All of this work is characterized by the same concern with real-life problems, careful consideration of past theory and research, meticulous analysis of data, and judicious conclusions that have become the hallmark of Gene's research.

Over the years Gene has contributed greatly to rural sociology through his undergraduate and graduate teaching. At the undergraduate level he has taught courses on the Structure and Function of Rural Society, Technological Change, the Rural Family, Eco-system Approach to Social Change, and Strategies for the Assessment of Social Change. His seminars on the Acceptance of Farm Practices, Technological Change, Developing Societies, Rural Development and Quality of Life have influenced the thinking and the careers of many sociology and rural sociology graduate students as well as those from extension education, ecology, agriculture journalism, and agricultural economics. He has probably had more influence on foreign students than anyone else in the department not only because his research and teaching has had great relevance to their societies but also because he has always been a patient, understanding and tolerant person who is more concerned with the professional development of his students than with his own aggrandizement. In his more than thirty years at Wisconsin, he has served as major professor to 43 students who have been awarded the Ph.D. Nearly half of these scholars are from foreign countries, including Canada, Germany, Spain, Colombia, Brazil, Egypt, Pakistan, the Philippines, and Korea. Almost all of them have returned to their home countries and have become important intellectual leaders either in their universities or in their governments. Many of his domestic students are professors in leading departments of rural sociology and sociology throughout the United States and Canada.

Gene has served his university and his profession well. He was Chairman of the Department of Rural Sociology during a critical period of its development and growth. He was Chairman of the University-wide Division of the Social Sciences, directed the Sociology of Economic Change program, and has served on numerous University and departmental committees. He has been President of the Rural Sociological Society and served as a member of the Council of the American Sociological Association. He is a member of the Sociological Research Association and several other professional organizations. He was a Fulbright Research Fellow in Australia, has been a Director of Research in Brazil for the Land Tenure Center and supervised students with the Sociology of Economics Change program in their study in Africa. He has participated in a number of international conferences including the planning of the First World Congress of Rural Sociology in Dijon, France in 1964. In 1981 he was selected for the Distinguished Rural Sociologist Award by the Rural Sociological Society for his many contributions to rural sociology.

In all of the years that I have known Gene, I have never seen him give vent to anger nor be unkind to any student or colleague. He has always treated his students as colleagues; many of them have co-authored papers with him. He has always been considerate of his students and helped them to achieve high standards of academic performance. In his quiet and modest way he has contributed greatly to sociology at Wisconsin and more than any other person is responsible for the pre-eminent reputation of Wisconsin's rural sociology program throughout the world. Best of all Gene has been a responsible and understanding colleague and friend to all of us at Wisconsin.

William H. Sewell
Vilas Research Professor of Sociology, Emeritus
University of Wisconsin–Madison

Record of Achievements

Personal Information

Born: July 7, 1916, Oak Ridge, Missouri

Family: Helena Emerson, Wife; Kenneth, Dean, and Karen, Children

Education: B.S., 1937, University of Missouri, Agricultural Economics
M.A., 1939, University of Missouri, Rural Sociology
Ph.D., 1949, University of Chicago, Sociology

Professional Appointments

1938–1940 Farm Security Administration, Rural Rehabilitation Supervisor

1940–1941 University of Missouri, Research Assistant in Rural Sociology and U.S.D.A., Agent with Bureau of Agricultural Economics

1942–1946 U.S. Army, Classification and Personnel Officer

1946–1949 North Carolina State College, Instructor in Rural Sociology

1949–1951 North Carolina State College, Assistant Professor in Rural Sociology

1951–1952 University of Wisconsin, Lecturer and Research Associate in Rural Sociology

1951– University of Wisconsin-Madison, Assistant Professor to
present Professor in Rural Sociology

1955–1959 Chairman, Department of Rural Sociology

1965–1969 University of Wisconsin, Director of Research in Brazil for Land Tenure Center

Honors and Awards

1948–1949 Social Science Research Council Fellow

1959–1960 Fulbright Research Grant, University of Melbourne, Australia

1981 Distinguished Rural Sociologist Award, Rural Sociological Society

Professional Activities

Major Achievements

An active member in the Rural Sociological Society during the years in which it was growing and defining its role. While President of the Society (1962–1963), he worked closely with a committee appointed to define the role of rural sociology in the U.S. and internationally. As the Society's representative on the Council of the American Sociological Society (later Association), 1964–1967, he was involved in the definition of its relationship to the ASA. He chaired the committee of the Rural Sociological Society on the Rural Environment (1972–1975) which later became a Section of the American Sociological Association. He co-chaired an Inter-American Research Symposium on the Role of Communication in Agricultural Development, in Mexico City, October 1964. He was also active in the North Central Regional Rural Sociology Committee sponsored by the Farm Foundation. From 1977 to 1981 he was editor of the Rural Sociology Monograph Series.

Professional Committees

1953–1962 North Central Regional Rural Sociology Committee Chairman, 1961

1953–1958 Diffusion Sub-Committee of NCR-5

1971–1973 North Central Strategy Committee for Commercial Agriculture

1963–1966 Development Committee of the Rural Sociological Society

1966–1969 Committee on Nominations, A.S.A.

1972–1975 Task Force on the Rural Environment, Rural Sociological Society, Chairman

1979–1981 Nominations Committee, Rural Sociological Society

Conferences

1964 Co-Chairman of Workshop on Goals and Values in Agricultural Development, Pullman, Washington

1964 Co-Chairman of First Inter-American Research Symposium of the Role of Communications in Agricultural Development, Mexico City

1964 Chairman of Workshop on Diffusion Research at First World Congress, Dijon, France

1968 Chairman, Workshop on Extension Research, Second World Congress in Netherlands

1976 Participant World Food Conference, Ames, Iowa

1976 Chairman, Seminar on Environment, Development and Quality of Life, Fourth World Congress for Rural Sociology, Torin, Poland

Professional Membership

Rural Sociological Society, Council 1960–1967
                          President 1962–1963
                          Book Review Editor 1954–1955

American Sociological Society, Fellow (Council 1964–1967)

Midwest Sociological Society

Society for Applied Anthropology

American Association of University Professors

Sociological Research Association

Wisconsin Sociological Society

Publications

Research Bulletins, Reports and Papers

1. *Planning for Family Relocation in the Wappapello Dam Basin Area* (with Cecil Gregory). Missouri Agricultural Experiment Station Bulletin 427, 1941.

2. *Acceptance of Improved Farm Practices in Three Coastal Plan Counties.* Technical Bulletin 98, North Carolina Agricultural Experiment Station, May, 1952.

3. "The Diffusion of Improved Farm Practices from Unit Test Demonstration Farms in TVA Watershed Counties of North Carolina" (with Frank A. Santopolo). Report to TVA Agricultural Relations Section, 1952.

4. *Adoption of Improved Farm Practices as Related to Family Factors.* Wisconsin Agricultural Experimentation Research Bulletin 183, December, 1953.

5. *How Farm People Accept New Ideas* (Joint with members of subcommittee for the Diffusion of Farm Practices). North Central Regional Publication No. 1 of the Agriculture Extension Series, Iowa State College, Ames, Iowa, 1955.

6. *Effectiveness of a Clothing Handbook in Teaching 4-H Club Members.* Extension Service Bulletin 522, July, 1956.

7. *The County Extension Agent in Wisconsin.* Research Bulletin 203, Agricultural Experiment Station, University of Wisconsin-Madison, September, 1957.

8. Preliminary Report No. 1–Roles of County Extension Agents. "Recruitment, Training and Selected Personal Characteristics of 90 Wisconsin County Extension Workers," Department of Rural Sociology, March, 1956.

9. Preliminary Report No. 2–Roles of County Extension Agents. "Role Discrepancy as a Measure of Stress in Cooperative Extension Work," Department of Rural Sociology, March, 1956.

10. Preliminary Report No. 3–Roles of County Extension Agents. "Clientele and Cooperating Associations and Agencies as Viewed by 90 Wisconsin County Extension Workers," Department of Rural Sociology, May, 1956.

11. Preliminary Report No. 4–Roles of County Extension Agents. "Program Planning, Local Leadership and County Coordination of 90 Wisconsin County Extension Workers," Department of Rural Sociology, September, 1956.

12. "Characteristics of Young Farm Families" (with F.C. Fliegel and James H. Copp). Preliminary Report No. 1 of "Social and Economic Aspects of a Farm and Home Development Program for Young Farm Families," Department of Rural Sociology, University of Wisconsin, 1956.

13. "Some Characteristics of Farm and Home Development in Wisconsin: As Viewed by the Agents" (with Charles L. Cleland). Preliminary Report No. 2, Department of Rural Sociology, University of Wisconsin, 1957.

14. *Goals in Farm Decision-Making As Related to Practice Adoption* (with Donald E. Johnson), Research Bulletin 228, Agricultural Experiment Station, University of Wisconsin-Madison, 1961.

15. *Five Years of Farm and Home Development in Wisconsin* (with Donald E. Johnson). Research Bulletin 228, Agricultural Experiment Station, University of Wisconsin-Madison, 1961.

16. *Aspirations, Work Roles and Decision-Making Patterns* (with Lakshmi K. Bharadwaj). Research Bulletin 266, Agricultural Experiment Station, University of Wisconsin-Madison, 1966.

17. *Rural Migrants in Central Brazil: A Study of Itumbiara, Goias* (with J.C. van Es, Senior author and Joao Bosco Pinto). Land Tenure Center Research Report No. 29, University of Wisconsin-Madison, 1968.

18. "Sociological Aspects of Colonization as Viewed from Brazil" (with Sugiyama Iutaka). Land Tenure Center Report No. 37, University of Wisconsin, June, 1967.

19. "A Agricultura e o Homen no Distrito Federal, Brasil: Relatorio Preliminar de uma Investigacao Sociologica " (Agriculture and Man in the Federal District of Brazil: Preliminary Report of a Sociological Investigation) (Jose Pastore and Fernando Rocha, Senior Authors), Research Report No. 28-P, Land Tenure Center, University of Wisconsin, March, 1968.

20. "Comparison of Migrants in Two Rural and an Urban Area of Central Brazil." Land Tenure Center Research Paper No. 35, February 1969.

21. "Stratification and Innovative Behavior: A Test of Cancian's Hypothesis" (with John Gartrell, Harley Presser, Jr.). Paper, Rural Sociological Society, San Francisco, 1969.

22. "University and Non-University Research." Paper, 7th European Congress for Rural Sociology, Munich, Germany, 1970.

23. "An Approach to the Study of Agricultural Systems." Paper, Rural Sociological Society, Washington, D.C., 1970.

24. *Wisconsin Incorporated Farms I: Types and Characteristics and Trends* (Richard Rodefeld, Senior Author). Department of Rural Sociology, University of Wisconsin-Madison, December, 1971.

25. *Wisconsin Incorporated Farms II: Characteristics of Resident Owners, Hired Managers and Hired Workers* (with Richard D. Rodefeld). Department of Rural Sociology, University of Wisconsin–Madison, December, 1971.

26. "A Systems Approach to Policy in Changing Agriculture." Paper, Third World Congress, Baton Rouge, Louisiana, 1972.

27. *Quality of Life in Kickapoo Valley Communities* (with P. Wopat, and J.G. Linn, C. Geisler and D. McGranahan). Report II, Institute for Environmental Studies, University of Wisconsin–Madison, September, 1973.

28. "Futurology and Quality of Life in Sociological Research." Paper presented at Rural Sociological Society Meeting, Montreal, Canada, August, 1974.

29. *Public Programs, Services and Policies in Four Northwestern Counties* (with Virginia Lambert, David McGranahan, Oscar B. Martinson). Report No. 41, Institute for Environmental Studies, University of Wisconsin–Madison, December 1974.

30. *The Use of Leader Ratings to Assess Community Services and Characteristics in the Kickapoo Valley* (with David McGranahan, Jon Huchison and Charles Geisler). Institute of Environmental Studies Report 45, University of Wisconsin–Madison, April, 1975.

31. *Social Determinants of Subjective Well-Being in Northwestern Wisconsin* (with Nancy Ahrens). Research Bulletin R2968, College of Agriculture, University of Wisconsin–Madison, December, 1978.

32. *"Community Growth, Outside Connections and Community Satisfaction in Northwestern Communities* (with David McGranahan). Paper contributed to International Sociological Association section on Rural Community, 1978.

33. "Involvement of Wives in Farm Tasks as Related to Characteristics of the Farm, the Family and Work Off the Farm" (with Nancy Ahrens). Paper presented at the Rural Sociological Society, Burlington, Vermont, August, 1979.

34. *Work Roles, Decision-making and Satisfactions of Farm Husbands and Wives in Wisconsin 1962 and 1979,* Research Report R3147, College of Agricultural and Life Sciences, University of Wisconsin–Madison, 1981.

35. "Family Farm Typology for the Analysis of Family Roles and Attitudes" (with Keith Moore). Paper presented at the Rural Sociological Society, Guelph, Ontario, August, 1981.

36. "Community Structure and Community Satisfaction in a Rural Region" (with David McGranahan). Paper presented at the Rural Sociological Society, Guelph, Ontario, August, 1981.

37. "Allocation of Family Labor On and Off the Farm: Its Conditions and Conse-
    quences" (with Keith Moore). Paper presented at the Rural Sociological Society,
    San Francisco, September, 1982.

Journal Articles and Chapters

1. "Why Don't Farmers Accept New Agricultural Practices," *North Carolina Agricul-
   turist*, 1949.

2. "A Socio-psychological Study of the Adoption of Improved Farming Practices,"
   Rural Sociology, 14 (March, 1949), 68–69.

3. "Sources of Information for Improved Farm Practices," *Rural Sociology*, 15 (March,
   1950), 19–30.

4. "A Socio-psychological Approach to the Study of the Acceptance of Innovations in
   Farming," *Rural Sociology*, 15 (December, 1950), 352–364.

5. "Sources of Information for Improved Farm Practices," *North Carolina Agricul-
   turist*, March, 1951.

6. "Social Isolation and Response of Farmers to Agricultural Programs," *American
   Sociological Review*, 16 (December, 1951), 836–837.

7. "Informal Leaders and Innovators in Farm Practices," *Rural Sociology*, 17 (September, 1952), 272–275.

8. "Change in Farm Technology as Related to Familism, Family Decision-Making, and
   Family Integration," *American Sociological Review*, 19 (February, 1954), 29–37.

9. "Techniques of Assessing Farm Family Values," *Rural Sociology*, 19 (March, 1954),
   39–49.

10. "The Role of Communicating Agents in Technological Change," *Social Forces*, 34
    (May, 1956), 362–367.

11. "Joint Decision-Making as a Function of Status and Role," *American Sociological
    Review*, 23 (April, 1958), 187–192.

12. "Consensus on Role Definition of County Extension Agents Between the Agents
    and Local Sponsoring Committee Members," *Rural Sociology*, 23 (June, 1958),
    184–197.

13. "An Introductory Note on the Social Aspects of Practice Adoption," *Rural Sociology*,
    23 (June, 1958), 97–102.

14. "Perception of Functions, Organizational Orientation and Role Definition of a Group of Special Extension Agents" (with Richard Smith), *The Midwest Sociologist*, 21 (December, 1958), 19–28.

15. "Communication and Technological Change in Rural Society " and "The Process of Acceptance of Innovations in Rural Society," Two chapters in *Rural Sociology*, A.L. Bertrand, ed., McGraw Hill Book Co., 1958.

16. "The Communication of Information on Innovations in Agriculture." In Wilbur Schramm, ed., *Studies of Innovation and Communication to the Public*, Institute for Communications Research, Stanford University, 1962.

17. "Human Factors in the Changing Rural Scene," *The Journal of the Australian Institute of Agricultural Science*, 26 (March, 1960), 11–18.

18. "Farmer Political Participation as Related to Socio-Economic Status and Perception of the Political Process" (with Ralph K. Huitt), *Rural Sociology*, 26 (December, 1961), 395–408.

19. "Communication and Acceptance of Recommended Farm Practices Among Dairy Farmers of Northern Victoria" (with Joan Tully and Hartley Presser), *Rural Sociology*, 27 (June, 1962), 116–197.

20. "The Diffusion of Technical Knowledge as an Instrument of Economic Development" (with William N. Parker, W. Paul Strassmann and Robert S. Merrill). Symposia Studies Series No. 13, the National Institute of Social and Behavioral Science, Washington, D.C., December, 1962.

21. "A Comparison of Husbands' and Wives' Responses with Respect to Who Makes Farm and Home Decisions" (Denton Morrison, Senior Author), *Marriage and Family Living*, 25(August, 1963), 349–351.

22. "Factors in Decision-Making in Farming Problems" (with Joan Tully and Hartley Presser), *Human Relations*, 17: 4 (November, 1964), 295–320.

23. "Some Perspectives on Change in Rural Societies," *Rural Sociology*, 29: 1 (March, 1964) 1–17.

24. "Alguns Problemas do Planejamento de Pesquisa Sobre Mudanca Social em Areas Rurais do Brasil," *Sociologia*, 27: 1 (March, 1965), 37–48.

25. "Decision-Making in German and American Farm Families: A Cross-Cultural Comparison" (with Eugen Lupri), *Sociologia Ruralis*, Vol. V., No. 4 (1965), 366–375.

26. *Rural Sociology*, 2500 word article in *Encyclopedia Britannica*, August, 1965.

27. "The Selection, Definition and Conceptualization of the Problems of Studies in Communications" in *Report of the First Inter-American Symposium on the Role of Information in Agricultural Development,* Delbert Myren (ed.), Mexico City, 1964.

28. "Dimensions of Aspirations, Work Roles, and Decision-Making of Farm Husbands and Wives in Wisconsin" (with Lakshmi K. Bharadwaj), *Journal of Marriage and the Family,* 29(November, 1967), 703–711.

29. "Aspirations and Attainments Among German Farm Families" (with J.C. van Es), *Rural Sociology,* 32(December, 1967), 446–455.

30. "Aspirations and Task Involvement as Related to Decision-Making Among Farm Husbands and Wives" (with Lakshmi K. Bharadwaj), *Rural Sociology,* 33(March, 1968), 30–45.

31. "The Role of the Extended Family in Migration and Adaptation in Brazil" (with Joao Bosco Pinto and Jose Pastore), *Journal of Marriage and the Family,* 30(November, 1968), 689–695.

32. "Some Problems of Development in the Central Plateau of Brazil," Chapter in *Sociology of Underdevelopment* by Carle C. Zimmerman and Richard E. DuWors (eds.), Toronto 2B, Ontario: Copp Clark Publishing Co., 1970.

33. "A Refinement of the Resource Theory of Family Power." Paper presented at the American Sociological Association, Boston, Massachusetts, August, 1968. *Sociological Focus,* Winter, 1968.

34. "Consensus in Aspirations for Farm Improvement and Adoption of Farm Practices" (with Sylvia Guerrero), *Rural Sociology,* 34(June, 1969):182–196.

35. "Response Stability in Survey Research: A Cross Cultural Comparison" (John C. van Es, Senior Author), *Rural Sociology,* 35 (June, 1970): 191–205.

36. "Interaction of Sociological and Ecological Variables Affecting Women's Satisfaction in Brasilia" (Suzanne Smith, Senior Author and Jose Pastore), *International Journal of Comparative Sociology,* 12:2(June, 1971):114–127.

37. "A Mulher e a Modernizacao de Familia Brasiliera" ("The Women and Modernization of Brazilian Family") (Majorie Gans, Senior Author and Jose Pastore), Pesquisa e Planejamento n. 12—Outubro de 1970. Revista do *Centro Regional de Pesquisas Educationais* "Prof. Queirox Filho"—Sao Paulo.

38. "Distance and Intergenerational Ties of Farm Families" (with Sylvia Guerrero and Spring Ginsberg), *Sociological Quarterly,* 13 (Summer, 1972): 383–396.

39. "Changing Structure of Family Farms in the United States," *Wies Wspolczsna* (Poland), 15 (Winter, 1971): 83–91.

40. "Canonical Analysis of Farm Satisfaction Data" (with Lakshmi Bharadwaj), *Rural Sociology*, 38(Summer, 1973): 159-173.

41. "Curvilinear and Linear Models Relating Status and Innovative Behavior: A Reassessment" (with J.W. Gartrell and Hartley Presser), *Rural Sociology*, 38(Winter, 1973): 391 – 411.

42. "Occupational Satisfaction of Farm Husbands and Wives" (with Lakshmi Bharadwaj), *Human Relations*, 26(October, 1974): 739-753.

43. "The Rural Environment: Quality and Conflicts in Land Use" (with Lowell Klessig). Chapter in *Rural U.S.A. Persistence and Change*, Thomas R. Ford, ed. Ames: Iowa State Univ. Press, 1978.

44. "Quality of Life: What is it?" (with Boguslaw Galeski), *Newsline*, January, 1977.

45. "Feelings of Powerlessness and Social Isolation among Large-Scale Farm Personnel" (with Oscar B. Martinson and Richard Rodefeld), *Rural Sociology*, 41 (Winter, 1976): 452-471.

46. "Ideology and Social Indicators of the Quality of Life" (with F.H. Buttel and Richard Rodefeld), *Rural Sociology*, 41 (Winter, 1976): 452-471.

47. "Outdoor Recreation and Environmental Concern: A Restudy" (with C.C. Giesler and Oscar B. Martinson), *Rural Sociology*, 42 (Summer, 1977): 241-249.

48. "Predicting Perceived Well-being" (with Lakshmi Bharadwaj), *Social Indicators Research* 4(1977): 421-439.

49. "Correlates of Subjective Well-being in Northern Wisconsin" (with David McGranahan), *Social Indicators Research*, 5 (1978): 211-234.

50. "Life, Domain Predictors of Satisfaction with Personal Efficacy" (with Lakshmi Bharadwaj), *Human Relations*, 33 (1980): 165-181.

51. "Farm Families and Family Farming" in R.T. Coward and W. Smith, *The Family in Rural Society*. Westview Press, 1981.

52. "Subjective Indicators and the Quality of Life," Chapter 16 in *Social Structure and Behavior: Essays in Honor of William H. Sewell*, New York: Academic Press, 1982.

53. "Religion and Community-Oriented Attitudes" (with Oscar B. Martinson and F.H. Buttel), *Journal for the Scientific Study of Religion*, 21:1 (March, 1982): 48-58.

Ph.D. Advisees:

1954	Martin L. Cohnstedt	University of Regina, Sasketchewan, Canada
1955	Frederick C. Fliegel	University of Illinois, Champaign–Urbana
1955	Thompson Peter Omari	UN Economic Commission on Africa, Ethiopia
1958	Charles L. Cleland	University of Tennessee, Knoxville
1958	James K. McDermott	USAID, Washington, D.C.
1958	Harry Schweitzer	Director of Extension, University of Illinois, Champaign–Urbana
1962	Denton E. Morrison	Michigan State University, East Lansing
1965	Gustavo Jimenez	Catholic University, Bogotá, Colombia
1965	Abdul Siddiqi	Peshawar University, Pakistan
1966	Ronald L. Johnson	West Virginia Institute of Technology, Montgomery
1966	Raymond D. Nashold	State of Wisconsin, Dept. of Health and Human Services, Madison
1966	Rajpal Rathie	Indiana State University, Terre Haute
1966	Donald E. Johnson	University of Wisconsin–Madison
1967	In Keun Wang	University of Seoul, Korea
1967	Joao Bosco Pinto	Inter-American Institute of Agricultural Science, Brazil
1967	Mohamed Gamel Rashed	University of Cairo, Egypt
1967	Herbert Smith	Western Michigan University, Kalamazoo
1967	Frank Leuthold	University of Tennessee, Knoxville
1967	Eugen Lupri	University of Calgary, Alberta, Canada
1968	Jose Pastore	Ministry of Labor, Brasilia
1968	Garfield Stock	Extension Business Outreach, University of Wisconsin
1968	Fernando Rocha	Ministry of Labor, Brasilia
1969	Johannes van Es	University of Illinois, Champaign–Urbana
1969	Howard D. Shiels	University of Wisconsin–La Crosse
1969	Lakshmi Bharadwaj	University of Wisconsin–Milwaukee
1969	James Hilander	Virginia Polytechnic Institute, Blacksburg

1972	Sylvia Guerrero	University of Philippines, Manila
1972	Erasmus Monu	University of Manitoba, Brandon, Canada
1972	Rodrigo Parra	University of Los Andes, Colombia
1973	Lowell Klessig	Environmental Resources, University of Wisconsin
1973	John Gartrell	University of Alberta, Canada
1974	Richard Rodefeld	Farmer, Cottage Grove, Wisconsin
1974	Susanne Saulniers	Ford Foundation, Nairobi, Kenya
1975	Ekong Edem Ekong	University of Ife, Nigeria
1977	E.G. Nadeau	Governor's Employment and Training, State of Wisconsin, Madison
1978	Oscar B. Martinson	University of Minnesota, St. Paul
1978	Jon K. Huchison	School of Architecture, University of Virginia, Charlottesville
1978	Carolyn Baylies	University of Leeds, England
1978	Eduardo Lopez-Aranguren	University of Madrid, Spain
1979	Charles Geisler	Cornell University, Ithaca, New York
1980	David McGranahan	Economic Research Service, USDA, Washington, D.C.
1980	Peter O. Agbonifo	University of Benin, Nigeria
1980	Harriet Moyer	University of Wisconsin Extension–Madison

Introduction

This volume of original essays was commissioned by the faculty of the Department of Rural Sociology, University of Wisconsin–Madison to be a celebration of Eugene A. Wilkening's three decades of creative scholarship, warm colleagueship and wisdom as a teacher. True to his character, he agreed to the idea of a Festschrift only if it could make a contribution to knowledge and resist the temptation to eulogize. Only time can measure the degree of our success.

Within the very broad sweep of social scientific concerns with technology and social change, Wilkening's work and professional interests provide an opportunity to explore four areas. These are: adoption and diffusion of innovations, the changing structures of agriculture and their relation to technological changes, public reactions to technological innovations and finally the effects of technological and social changes on rural life: the farm family, the role of women in agriculture, social class formations in rural areas, community power structures, and citizen mobilization. What we have attempted to do in the Festschrift, therefore, is to devote attention to each of these four areas. While they are selective, to be sure, they have been, are now, and surely will continue to be significant areas of sociological inquiry.

In *Technology and Social Change in Rural Areas* we have used the history of technological innovations which have profoundly altered rural societies since World War II as the vehicle for the organization and flow of the book. There is a second obvious and compelling reason for this choice. Gene Wilkening has been a major analyst and interpreter of these events as they occurred and thus there exists a close relationship between his career and the period of history he observed and reflected upon.

This book enters the stream of history at the point where great efforts were being made to encourage the processes of adoption and diffusion of technological innovations. The partial successes of these efforts dramatically reshaped the structures of agriculture and ultimately the social order of rural segments of society, even total societies. The export of modern technology to developing nations resulted in successes and failures, some dramatic. More recently, nuclear accidents, atomic warfare, genetic engineering and other technological innovations with potentially hazardous effects on the natural environment and humans have created demands for a more humane application of technology.

The authors were asked to reflect on these experiences and to challenge social scientists with an interpretation that goes beyond a state-of-the-art review or catalog of needed research. But before introducing specific essays, it will be helpful to expose some of the undercurrents which flowed through the events of these thirty years, and are visible in the essays which follow. Calling attention to these diffuse but pervasive forces which mingle with technological development and social changes will clarify and enlighten the discussions. The authors were not asked to deal with these particular forces, but neither were they instructed to ignore them. In consequence, some have drawn them openly into their discussions and others have left them in the wings.

Undercurrents of Change

Technology and social change are related forces in history. Perhaps it is more customary to think of technology as the causal factor and social change as the consequence than to view them from the reverse side. Technological determinism has been with us a long time, wearing a variety of masks. But there are those societal observers who insist that social changes are stimulated by forces in addition to technology. Therefore social structures have a degree of independence which permits them to exert reciprocal influences on technology and to influence its creation, adoption, diffusion and its effects on the future course of social history. This condition of partial dependence permits reciprocal or alternating causality to occur and creates an ambiguous situation which provides observers, analysts, and interpreters the freedom of a Rorschach ink blot. Interpretations are guided as much by the interpreter's own values, philosophical assumptions and ideological dogma as by empirical reality. There is, then, an apparent indeterminancy in the relation between technology and social structures which has generated vacillation and debate about the relative importance of technology and social change in history.

Technology is a unique combination of capital and labor designed purposely to enhance the creation of a desired product. While the term is used also to refer collectively to the summary or dominant mode of these combinations in a period of history, it is the more circumscribed sense in which social scientists grapple with technology in relation to social change. Thus, any product is a function of capital, labor, and a technology which combines them to form the product. Institutions are the "rules of the game" governing acceptable uses of capital and labor. Furthermore, capital can be usefully regarded as consisting of (1) natural resources and (2) man-made resources – the physical products of past and current technologies. Similarly, labor is divisable. It consists of the number of persons available for work but also treats independently those differences in demographic characteristics, such as age and sex. Also, skills, experience, knowledge, and motivations which affect ability or willingness to work are analytically distinct dimensions. Technology is the creative blending of these ingredients for productive efforts.

To be sure, this is an extremely generalized scheme, but because of its simplicity, it is widely applicable and may encourage the discovery of insights. For example, the technological innovations in U.S. agriculture during the mid-twentieth century occurred largely within a single matrix of institutions at a particular point in time. Consequently, few persons involved in technological innovations regarded institutions as a factor in reckoning rates of adoption and diffusion or consequences of innovations. Therefore, institutions were either ignored or left as implicit components in the new technologies. But when these new technologies, including implicit elements, were exported to other nations with strikingly different institutions, failures occurred. Then, and only then, did the trial and error approach reveal experientially what was already presumed in the generalized scheme; namely, institutions are critical ingredients in the development of technology and its implementation.

Capital, labor and technology are owned, or at least their uses controlled, by persons representing either themselves or some interested party. Again, social analysts often have overlooked institutions as items which can be controlled and/or used in combination with capital and labor elements. Yet, they are used as integral components in technological innovations; usually implicitly but occasionally, explicitly. It is the value of institutions and the purposeful control and/or use of them that has been a focal point of critical analyses of societies. Beginning with the work of a radical structural-functionalist, Floyd Hunter and a radical Weberian, C. Wright Mills, concern with the control of institutions and its importance in history has moved to the present fractured Marxist debate among "instrumentalists" and "structuralists." Thus, the issue of control and use of institutions is a strong undercurrent.

There is another related but separable current which turns on the issue of control of decision-making especially regarding public investment policies. It is most visible in the demands of citizens for greater participation in the making of decisions which affect their lives. Explanations for its widespread appearance, multi-national in scope, read like a litany of basic sociology: specialization in the division of labor, bureaucratization, centralization of decision-making, declining importance of locality as a basis for social organization, urbanization and alienation. Whatever its true source, or sources, the demand for increased citizen participation became a significant social force with active proponents in government and the citizenry.

Within government, there has been a concern with regaining and retaining public support which is manifested in requirements for citizen participation in virtually all government assistance programs. Bureaucrats were quick to recognize that citizen members of committees and citizen advisory groups could be co-opted and transformed into advocates for their programs. In many instances, government agency sponsored citizen committees, public hearings, and forums became transparently ritualistic and self-serving. Consequently, the surge for citizen initiated involvement became an even stronger force for citizen participation.

Advocacy groups often have pursued single issue political pressure tactics and frequently have drawn their members from a very small segment of society. However, with strong strategy, financial backing, good organization and coalition formation, citizen initiated involvement groups have been effective in getting their demands heard. At times their efforts have coalesced into social movements, as is evident in the civil rights movement, the anti-war movement, the environmental movement, or the women's liberation movement.

It appears there has been a questioning of primary standards of system performance in these outcroppings of insistent citizen participation: equity and efficiency in the allocation of public resources. In only a few instances has the question of public versus private ownership of resources surfaced. Technology has been the focal issue for only a few of these advocacy groups, but it has been a peripheral issue for many. In other instances the substantive issue has been equity in the distribution of benefits and costs among various segments of the public. Seemingly, equity in the distribution of benefits has had more currency than equity in the distribution of costs. Both benefits and costs go beyond financial considerations to include health and safety, freedom of choice, access to public resources, rights of unborn generations, and other expressions of valued properties — tangible and intangible.

The equity question became a matter of public concern at a time when the exhaustability of natural resources came to the public consciousness and when increasing competition in international trade created doubts about the infinite expandability of markets. The limitlessness of these two capacities was for many years a basic assumption of the growth model of national and international development. In that scheme equity was to be achieved by means of increased efficiency, growth in productivity, increased sales, and profits which would be re-invested to continue the process.

Efficiency, lower cost per unit of output, also was a primary standard of system performance. The usual method of achieving efficiency was to create new combinations of capital and labor — technology — which reduced the quantity of labor needed to maintain or perhaps increase the level of productivity. Customarily, we think of these creations as "labor-saving" technologies, which they were in many instances. Capital, in one or more of its forms, was substituted for quantities of labor. But one should also notice that numbers of workers is only one aspect of the labor component of production. Trade-offs can be made among the aspects of labor to create new technologies having greater efficiency. For example, mechanical devices or reorganization of work flow, as in the assembly-line factory, have been used to increase efficiency *and* productivity without a decrease in numbers of workers. Or as a second example, workers willing to work for less compensation may be substituted for higher paid workers resulting in greater efficiency with no loss of productivity. Labor manipulation notwithstanding, it is true that many industries were able to increase efficiency and productivity by emphasizing the use of natural resources and their derivatives and by increasing investments in man-made resources. To the advocates of the growth model and the efficiency standard, recent developments in bioengineering open a window onto a vast new frontier of industrial and commercial opportunities.

How future advocates of equity will fare against efficiency adherents remains to be seen. But clearly the struggle between these two basic standards of system performance has been a major undercurrent the past thirty years and surely will continue for some distance into the future.

During these same three decades the major industrial nations have been undergoing fundamental restructuring of their national economies. It has been widely assumed that human welfare could best be enhanced by real increases in the level of income which people could use to purchase the goods and services they desired. Furthermore, since income is a reward for work, job creation was viewed as the proper means of achieving more real income. The traditional sources of employment and income growth have been (1) increased productivity in agriculture, (2) greater exploitation of natural resources and (3) improved efficiency and productivity in manufactured products; especially those for export. In the years since World War II the extractive and goods-producing industries have ceased to be the growth mode of national economies, in either job or income. Certainly they remain as critical components of the economy. But the source of growth has shifted to the non-goods producing industries. This is true in every industrially developed nation, regardless of its system of political economy.

A key element in "deindustrialization" is the migration of capital, especially finance capital, from one industry to another, from one region to another and from one nation to another. The latter movement is associated particularly with multinational corporations, themselves a product of post-War technologies. Plant closures are a predictable consequence of capital mobility and are accompanied by hardships to workers and communities suffering the loss of an establishment. The undercurrent is the free movement of capital and the extent to which job security and community stability should be allowed to place constraints upon it. While this issue is only now threatening to become a matter of public debate in the U.S., the forces at work extend back at least thirty years. And indeed, the issue is already a point of public debate in most industrial nations of Europe.

The Essays

Four sets of essays were commissioned by the Festschrift Committee to coincide with the four dominant concerns of Professor Wilkening, as noted above. The order of their presentation recapitulates the progression of his attention to these topics in the course of his career and does not imply that events and issues which dominated the early years are any less important today; it merely reflects the order in which their significance emerges in the research and theoretical writings of Professor Wilkening. It may suggest to some readers an implied causal order in historical events such that rural social changes are produced solely by innovations in technology, especially those applied to production agriculture. Such an interpretation is not intended and it would contradict Professor Wilkening's rejection of technological determinism.

The first set of essays deals with issues involved in the adoption and diffusion of technological innovations in agriculture. Fliegel and van Es review the record of research based on a paradigm in which the individual farmer was treated as the key unit of analysis. Recent critics maintain that such a focus fails to deal adequately with institutional factors. Following their appraisal of competing paradigms, Fliegel and van Es offer some suggestions for a nascent paradigm which attempts to deal effectively with the full range of variables pertinent to adoption and diffusion of innovation.

In the second essay, Mazur maintains that public resistance to technological innovations, even resistance within the scientific community, is not a new phenomenon. By reviewing the historical record of resistance efforts, he concludes there are systematic conditions which presage such public actions. Moreover, resistance movements follow routine patterns of emergence. These are illustrated with details from the anti-flouridation and anti-nuclear resistance efforts. A key element in the progress of protest, according to Mazur, is the ebb and flow of larger national concerns having a related ideology. This thesis is examined using national polling data. By focusing on resistance, this essay strikes a useful balance with adoption and diffusion models which emphasize the conditions associated with positive public reactions to technological innovation.

To conclude this set of essays, the Festschrift Committee asked Professor Krimsky, a philosopher, to reflect on recent developments in biotechnology, human engineering and other aspects of high technology and to challenge the collective conscience of the social science community. It was our presumption that future issues surrounding technological innovations and their adoption and diffusion must include these topics. The fundamental question raised by Professor Krimsky is whether existing social institutions are adequate to control the development and uses of high technology. Taken seriously, this query poses both a moral and a scientific challenge to social science researchers.

The second set of essays deals with the consequences that technological innovations have had on the structure of agriculture. Dorner and Buttel examine the U.S. experience and Newby and Galeski focus attention on the "agrarian question" in European settings. The authors of these papers were asked to presume the historical reality of technological innovations in agriculture and to deal with the implications for the structure of agriculture as a basic industry, including the associated institutional matrix.

After reviewing the mechanical and chemical transformations of U.S. agriculture, and their many implications, Peter Dorner concludes, "There is indeed much to be critical about." But he does not stop with a litany of criticism. Rather he sees hope in new directions in science and technology. One need not endorse all technological innovations to reach the conclusion that basic technological developments have had a major positive impact on the output of food. Thus, he is optimistic about the future relationship of technology and U.S. agriculture, but with the admonition that potential and unintended detrimental impacts must receive as much attention as the fostering and diffusion of new technology.

One of the clear effects of widespread adoption of mechanical and biochemical innovations in U.S. agriculture has been the fall of the family farm. The drive for efficiency has lead to larger and fewer farms, often involving non-family labor and corporate ownership. The often heard lament for the family farm is challenged by Buttel who argues that the mourning should be for the demise of democracy in the organization of agriculture. Small family farms were a democratic alternative to "grossly unequal landholding systems of Western Europe and to the 'peculiar institution' of the antebellum Confederacy and its legacy in the sharecropper/crop lien system of the post-bellum South." He concludes that alternatives to family farms are needed which preserve the goal of democratization within the structure of agriculture; technological innovations in agriculture have exposed the means-to-an-end nature of the family farm and challenge the architects of society to find suitable substitutes.

In a fundamental sense finding alternatives to the extreme choices of either large estates or peasant small holdings while preserving democracy in an agricultural production system consistent with other production systems of society is the agrarian question. Newby explores the central position of landholding in the sociological analysis of agriculture by re-thinking nineteenth century theorists: Marx, Weber and Kautsky. He concludes that while they do not adequately furnish the answers, they do point properly to the kind of questions which should be asked. Moreover, "rural sociologists must direct far more attention to landholding in order to establish a distinctive sociology of agriculture."

The Polish experience in grappling with the agrarian question provides an unusual opportunity to examine several landholding systems. Galeski has been an interested observer of the Communist Party's efforts to find an acceptable answer to the agrarian question. Using agricultural output as the primary criteria of success, he argues that centralized planning of production and state ownership of land combined with a state monopoly in the marketplace have been the least effective solution in Poland. Private ownership of the means of agricultural production and the operation of uncontrolled, or partially controlled, market forces provide a better solution to the agrarian question.

Technology has profoundly influenced patterns of living and the quality of life in rural areas; traditional roles have been altered, and new ones introduced; old institutions have given way to newly formalized relationships among people and organizations. Selected aspects of social changes in rural life are discussed in the next section by three authors who have worked with Professor Wilkening in his analysis of the impacts of technology on rural life.

Gartrell discusses the potential effects of technology transfer on rural communities in developed and developing nations. The chapter begins with an assessment of recent efforts to understand how technology transfer in agriculture affects rural communities. He argues that such transfers often lead to dependency which results in uneven development and affects community autonomy and inequality.

Technological innovations in U.S. agriculture have dramatically reduced the number of farmers and farm workers. Consequently, rural life is no longer organized primarily around agriculture as it once was. McGranahan investigates the effects of these changes on the social and spatial relations or rural communities in the U.S. The analysis is a peruasive case for the idea that some local interests are associated with residence in the community while others are economic and originate in the spatial segmentation of markets. Thus, he concludes that rural communities and their changing structure cannot be understood without reference to social class and economic interests.

The chapter by Haney focusses on the role of women in the nearly three-fourths of the U.S. agricultural production units classified as marginal which includes a large proportion of the family farms, one of the central institutions in rural life. In spite of the critical role of women in it, the conjunction of women, work and the family is an area of social reality with a very limited research history. Yet, it is clear that women are principal actors in the farming system.

In the final set of essays, Morrison, Schnaiberg and Thiesenhusen examine the question of whether technology can be molded to fit *existing* institutional arrangements and the supply of factor inputs, capital and labor. In the first of these essays, Morrison examines the contemporary debate over whether certain technological styles are "appropriate" or "inappropriate." There is an active social movement which argues that modern, industrial technologies have undesirable impacts on equity, the physical environment and natural resources, and the quality of life. The major features of the movement and its social origins are examined, then the emerging high technology is characterized and analyzed in terms of impacts which concern the appropriate technology movement.

Schnaiberg's essay is a sharp and skillful assessment of the political efficacy of the appropriate technology movement, particularly in the U.S. He argues that the failure of the movement resides in its assumption that *ideas* rather than *interests* shape the social course of productive technology. The movement has not responded effectively to the resistance of existing production groups and the state, nor has it successfully anticipated and defended against the co-optation of the movement by such groups. To alter these limits the movement would have to incorporate a much more explicit conflict orientation and accommodate a greater diversity of political strategies.

In the final essay, Thiesenhusen relates the concept of appropriate technology to the adoption and diffusion of U.S. agricultural technology in less developed countries. After pointing to the successes of the utilization technology in U.S. agriculture, he shows that highly skewed distributions of income and wealth (landholdings) and high incidence of unemployment makes the institutional system in most LDCs much different from what prevailed in the U.S. while it was industrializing. He then argues that the problems LDCs confront are at least as much social as technological. Finally, he concludes that U.S.-type institutional structures and technologies usually are not appropriate for agricultural development in emerging nations, at least not without extensive adaptation.

Gene F. Summers
Professor of Rural Sociology
University of Wisconsin–Madison

Part 1

Adoption and Diffusion

Chapter 1

The Diffusion-Adoption Process in Agriculture: Changes in Technology and Changing Paradigms

Frederick C. Fliegel and J.C. van Es

Introduction

The great technological and organizational transformation of United States agriculture during the middle third of the twentieth century is reflected in a unique way in the research efforts of rural sociologists. Starting in the late 1930s, rural sociologists' research on the diffusion and adoption of agricultural technology grew rapidly for several decades, reached its zenith around 1960, and has declined substantially since that date. During this period, research focused on the process of introducing new technology at the farm level. As the agricultural transformation took its course, the urgency of promoting rapid introduction of new technology to American agriculture largely disappeared, and the interest in diffusion research waned likewise.

In the second half of the 1970s, there has been a minor resurgence of interest in the diffusion and adoption research. However, as we will note later, the focus of this research has changed substantially from earlier preoccupations. The concern with technology currently is less with the process of diffusion and adoption of new technology at the farm level, than it is with ways in which one can deal with the consequences of the widespread adoption of that technology. While the

We wish to thank several reviewers of an earlier draft of this chapter for their useful comments: J.E. Kivlin, E.M. Rogers, A.W. van den Ban, and especially Gene F. Summers. Any remaining weaknesses are our responsibility, however.

earlier interest was in matters like the adoption of 2,4D weed spray, the concern currently is with programs that might succeed in dealing with the environmental consequences of the use of chemicals in agriculture.

The diffusion-adoption research tradition not only shows remarkable synchronization with the technological transformation of U.S. agriculture, it also strongly mirrors the intellectual traditions current within sociology during the period of inception and subsequent growth of the diffusion research tradition. The nature of the diffusion research tradition cannot be understood without an appreciation of its intellectual foundation: problem statement, research approach, and methodological strategy were dictated almost entirely by a contemporary intellectual tradition which defined sociological phenomena at the micro level, and explained behavior in terms of individual acts. Briefly stated, the diffusion/adoption research tradition viewed the farmer as actor, in a farm and local community situation, responding to stimuli concerning what were unquestioningly viewed as improvements in agricultural technology. The research tradition cannot be appreciated without acknowledging its debt to theoreticians like Ogburn, and without appreciating the tremendous intellectual ferment created by the methodological breakthroughs attributable to such pioneers as Stuart Chapin, Likert, and Guttman. As the behaviorist approach grew in strength, so did diffusion research. When this paradigm or combination of paradigms was challenged by alternative approaches to sociological inquiry, the support for diffusion-adoption research within the discipline also declined precipitously.

The discussion in this chapter is intended to provide an overview of the various efforts rural sociologists have made to focus on agricultural technology, first as a dependent variable in the context of identifying antecedents of adoption or nonadoption, and later as an independent variable, with emphasis on consequences. The early work is described as generally reflecting a concern with measurement: at what level of abstraction should agricultural technology be specified in order to generalize about the antecedents of adoption behavior? Can meaningful types of technology be identified to gain insight into diffusion processes? As the interest in antecedents of adoption faded for a variety of reasons, interest in differentiating agricultural technology into component types declined as well. But questions about consequences of agricultural technology gained in salience, and a research focus on certain kinds of technology (e.g. agricultural chemicals) has gained new prominence. The latter focus is relatively new, not yet well defined, and by and large does not build on earlier efforts to disaggregate agricultural technology into component types. The current focus on some kinds of technology rather than others appears opportunistic, stemming from concern with particular consequences rather than a concern with the broader diffusion-adoption process or technological change processes. As one might expect, a fair amount of both technical and conceptual confusion has been the result, and the chapter ends with suggestions for bringing some order to future research activities to reduce that confusion.

Diffusion-adoption research: The first generation

It is inevitably somewhat arbitrary to identify the origins of a particular body of work and thought. In the present case we are inclined to be pragmatic, if somewhat arbitrary, and designate the studies done at Iowa State University around 1940 as the starting point for a large volume of research on the diffusion of agricultural innovations. The original focus was on a single innovation, hybrid corn (Ryan and Gross, 1943), and it is relevant to note that the work was done in a state where agriculture is quite important, and the research concerned a major agricultural commodity, corn. Later on, the linkage between technological change in agriculture and individual profit came to be treated as a basic and often unstated assumption of diffusion research.

Research on diffusion of innovations was dominated, at least in the early years, by North American rural sociologists, and particularly those in the Midwest. Commercial agriculture is singularly important in the Midwest, it involves relatively large numbers of farm operators, and it is supported by a highly developed research and service infrastructure. The development of commercially viable corn hybrids was by no means a short-term endeavor but by 1940 some considerable success had already been achieved in realizing the great potential of the new varieties in farmers' fields. At the same time, as the new technology became available, concern arose about the willingness of farmers to adopt such innovations expeditiously. Policy makers, and also commercial interests, were frustrated in their efforts to quickly introduce new technology among the farm population. Questions arose about farmers' willingness to change, and ways in which the infrastructure might be altered or manipulated to expedite the adoption of new technology at the farm level. It is fair to say that the early research on the diffusion of hybrid corn and its adoption by farmers was noticed and emulated because of the very considerable economic potential of the innovation itself for a major crop and thus for the profitability of the industry. The stage was set for a concern with technological change as an agent of economic growth.

If the economic potential of hybrid corn is a hallmark of the early Iowa State research on diffusion, one must also recognize the sociological sophistication involved in the early studies. From a vantage point some 40 years after the work was done it is easy to see the outlines of what later became known as the "diffusion model" in that early work. That is, the characteristics of the farmer as actor, in a certain situation, with identifiable linkages to particular institutions and agencies, form the core of the work, and the several variables were analyzed singly and jointly to shed light on the individual adoption decision. The issue of profitability as a factor in that decision was more or less taken for granted. This assumption becomes important from the same 40-years-later time perspective, when some of the more controversial consequences of technological change in agriculture have become visible.

Concern about generalizing from the work on diffusion of hybrid corn is quite evident in the Iowa State studies. Such concern was expressed in a variety of ways. First, questions about the antecedents of adopting hybrid corn were put to different samples of farmers (*e.g.* Ryan, 1948), to determine whether the same patterns were evident. Second, a study of the antecedents of a different innovation was undertaken (Gross, 1949), and this was later expanded to an analysis involving ten innovations (Gross and Taves, 1952). In other words, a very direct effort was made to compare the correlates of adoption of one innovation with the correlates of adoption of a range of other innovations. And third, the distinguishing characteristics of hybrid corn itself were listed and discussed (*e.g.* Ryan and Gross, 1950), and similar efforts were made to specify the distinctive features of other innovations under study. In short, the foundation was laid for sorting out types of innovations from the larger and rather amorphous category, improved agricultural technology.

Indexes of adoption of innovations came into widespread use as the research interest in the diffusion process itself diffused to other research settings and individuals. Merging a variety of innovations into a single list, and computing individual farmers' adoption scores with reference to that list, was a direct response to the desire to generalize from one adoption situation to the next. In line with both policy preoccupation with the rapid diffusion of existing new technology, and the individual-actor emphasis of the theoretical orientation, the research focus in the 1950s and 1960s was on the general question: why do farmers adopt or not adopt agricultural innovations? Explicit interest in the distinguishing characteristics of particular innovations, and efforts to discern types of agricultural technology, were not vigorously pursued. Instead, diffusion researchers devoted considerable effort to describing the universe of improved technology and measuring an abstract and variable propensity among farmers to adopt that technology. Research interest shifted from at least an incipient concern with the nature of technology and efforts to specify particular types of technology, toward an interest in measuring a socio-psychological posture toward improved technology as a whole.

Concern about types of technology was not entirely submerged in the general effort to construct reliable and valid, multi-item adoption indexes. One example of the continuing effort to specify types of technology is the distinction between *improvements* and *innovations* (Wilkening, 1954). In retrospect, the effort to distinguish between improvements (modifications of existing technology) and innovations (departures from existing technology) can be viewed as part of the wider concern with delimiting the universe of improved technology. The distinction between improvements and innovations was intended as a means to arrive at better measures of the abstract socio-psychological dimension, a variable propensity among individual farmers to accept that which is new. The analytic emphasis was on sorting out the antecedents of adoption, with subsequent comparison of those antecedents most useful in explaining variability in the *several* measures of the dependent, adoption variable.

The distinctive characteristics of particular innovations also continued to attract at least passing attention during the 1950s, and somewhat more systematic attention in the 1960s. Both Lionberger (1960), and Rogers (1962), in their compendia of research on diffusion of innovations, paid attention to distinctive characteristics of innovations which were presumed to be influential in diffusion. As already noted, the emphasis on characteristics of innovations was directed toward explaining diffusion and individual adoption from a socio-psychological perspective. It was thus logical to attempt to specify characteristics of the innovation stimulus which could be expected to trigger the desired adoption response. Other criteria for defining the characteristics of technology are the consequences of the adoption of technology, or the nature of technology as such, but the emphasis of the 1960s was on the innovation as stimulus. Perhaps the most likely characteristic of agricultural innovations to be singled out was the fact that some innovations could be tried out on a small scale, before full adoption, while others could not. Divisibility for trial was expected to contribute positively to eventual adoption by permitting the farmer to gain experience with the innovation, while minimizing any risk inherent in trying something new. Trial on a small scale was viewed as a logical step in the individual learning process which culminated in adoption. A decade later, when farmers' *ability* to adopt innovations came to be questioned (Galjart, 1971), the same characteristic, divisibility, came to be viewed from a structural perspective under the heading of what economists call "scale neutrality." From the latter perspective, an infinitely divisible innovation could at least be tried by the individual farmer regardless of that farmer's scale of operations. Full adoption might well pose capital requirements beyond the farmer's means, however, and the presumed learning experience might not lead to adoption in the face of a structural constraint.

Efforts to specify relevant characteristics, or attributes, of agricultural innovations can be summarized by the early work of Kivlin (1960), and some of the studies that grew out of that work (*e.g.* Fliegel and Kivlin, 1966). The basic approach was a measurement approach, attempting to address the practical problem that antecedents of adoption of a given innovation were demonstrably imperfect in predicting the adoption of other innovations. Rather than, or in addition to, further specification and test of antecedents of adoption, defined as a readiness to accept the novel, it was argued that the degree to which a range of innovations possessed certain characteristics needed to be specified. Specification of novelty thus gave way to specification of a range of presumably salient characteristics. Innovations might be ranked, for example, on the degree to which they could be tried on a small scale before full adoption. Similarly, innovations could be ranked on their cost to the farmer, their potential for increasing returns, and so on. Such a procedure was intended to produce specialized adoption indexes, involving innovations having certain characteristics in common. With specialized indexes in hand, antecedents of adoption of particular types of innovation could be explored, with later comparison across the several types of innovations.

The work on characteristics of innovations can be viewed as a logical exten-
tion of earlier work, in which the characteristics of hybrid corn and other innovations were discussed as possibly influencing the diffusion process in distinct ways.
The early studies could do no more than suggest possibilities for systematic tests,
however. Specialized indexes, such as those mentioned above, represent an
advance in differentiating agricultural technology into potentially meaningful
types. Willingness to adopt certain innovations, and thus diffusion rates, were
hypothesized to be determined in part by systematic differences among innovations. All of that work was primarily directed toward improved specification of
the measure of adoption. Classifications were developed empirically, without
much attention given to developing a classification of technology related to a
theoretically grounded definition of technology. Nor was there serious questioning of the ends to be served by adoption; individual benefit was largely taken for
granted. The substantially socio-psychological approach to explaining the diffusion process remained basically the same throughout the period described to this
point, and the measurement issues involved were pursued but only partially
resolved.

Failure to resolve the measurement issues was at least partially due to a
decline in research interest in the diffusion process, discussed in the next section.
Following that discussion we describe a partial revival of interest, but from a
different perspective, in diffusion phenomena, and in types of agricultural technology.

A decline in adoption research: Some reasons for it

Though we have not made a tabulation of trends in budget and personnel allocations to research on diffusion phenomena we are confident that there has been a
decline since the mid-1960s among students of agriculture. Research on diffusion
of other than agricultural innovations flourished (see Rogers with Shoemaker,
1971), but that extension of the research tradition is not of direct concern here.
Several reasons for a decline in interest among rural sociologists are presented
below. They are intended as candidates for consideration rather than as the definitive explanation of a phenomenon.

Farm-level adoption is no longer viewed as problematic in the industrial
nations of the world. Agricultural production has in fact been radically transformed by the substantial absorption of a host of mechanical, chemical, and
genetic as well as corresponding managerial innovations. The displacement of a
substantial amount of farm labor has been an important (and probably initially
unanticipated) part of that transformation, as has the gradual increase in scale of
operation of the remaining farm units. Land and capital resources are concentrated in fewer and more highly specialized units than was the case a generation
ago, and these units are managed by individuals who are much better trained

than their predecessors. Contemporary farmers, as a group, have communication facilities available to them which are unprecedented in history. More and more agriculture has become intertwined with and inseparable from industrial society as a whole. The current technology of production is by no means static, but in the industrialized nations, the physical and social context into which innovations are now introduced is utterly different from that of 40 years ago. Earlier, it was the *failure* of farmers to quickly accept improved technology that stimulated research. Such failure was widely perceived as a societal problem. That is no longer the case. The generic issues underlying technological and social change processes have not necessarily changed, but consensus on what is to be defined as a "social problem" and thus worthy of research has changed. Currently it is the consequences of technological change, rather than any failure to accept such change, which are viewed as a social problem.

Export of the diffusion model to third-world agriculture revealed shortcomings in the approach. A certain amount of disenchantment with diffusion research, and thus also a decline in interest in specifying types of agricultural technology, stems from simply transplanting what is loosely called the "diffusion model" into other cultural contexts (Goss, 1979). The spread of diffusion research to other countries was at least in part another manifestation of the researchers' aspirations to a universalistic model for explaining adoption behavior. The initial cross-cultural tests took place mainly in Europe and Australia (*e.g.* Wilkening, *et al.,* 1962), and this research generally supported the U.S. findings. These supporting findings tended to enhance the stature of the diffusion model. However, when the diffusion model was applied in developing countries, especially in Latin America, the model was frequently found to be inappropriate. The hoped for agricultural transformation did not come about, and probably could not come about, as a result of the introduction of new technology in agriculture, because individual farmers could not be assumed to be free to act in their own interests.

Agricultural production in the third world often lacks the infrastructural support and the access to capital which was largely taken for granted in early diffusion studies. Where attitude toward credit was viewed as a relevant variable in the early studies, the sheer availability of credit becomes a more important constraint in third-world diffusion studies. The preceding illustration, concerning credit, can be used to point to a more basic shift in perspective, a gradual shift in theoretical posture away from the socio-psychological approach to diffusion, concerned with characteristics of the individual farmer, toward a focus on structural factors, or constraints. The essentially voluntarist diffusion model, based on an idealist/voluntarist theory (Flinn, 1982:3) which was quite pervasive in the rural sociology of that era, was challenged and found lacking. In the course of time it became evident that such factors as access to land, markets, and legal protection were major constraints in the modernization of agriculture in some third-world settings. North American diffusion researchers had taken the existence of a supportive infrastructure in their own setting for granted, and they were sometimes not sensitive to the consequences of considerably greater inequality in distribution of resources for the diffusion of innovations. Again, structural constraints came to be regarded as more salient and a conceptual shift away from the socio-psychological model was called for.

A dependence on survey methods is also implicated in the decline in diffusion research. A behaviorist approach and its methodological corollary, survey research, served well to specify characteristics of individual decision-makers, their access to information, the nature of the decision-making process, and farm level impediments to adoption. Such an approach continues to serve well, as a matter of fact, in many developing countries, where a local situation can be analyzed and useful information can be provided to assist action agencies in promoting acceptance of improved technology. Many such highly localized, very practical surveys continue to be carried out.

However, where structural factors are of major consequences as causes of non-adoption it appears that behavioristic models and survey techniques do not lend themselves to research on the kinds of questions being raised. To put it differently, research on macro-level problems is not well served by micro-level techniques. Questions about the functioning of credit institutions and the credit system, for example, are not always amenable to the conventional sample survey. The point here is that the research tradition is to some extent a victim of the techniques which brought it success: familiarity with and a preference for survey techniques among many researchers may actually work against making progress in research on contemporary diffusion problems.

In summary, several reasons have been cited for a general decline in research emphasis on diffusion-adoption phenomena. Lack of progress in specifying types of technology and pursuing research themes related to certain types of technology is part and parcel of that decline. In the following section we describe a resurgence of interest in some types of technology. This resurgence is a relatively recent phenomenon, it is clearly linked with the older diffusion research tradition, but for the most part different questions are being raised and different approaches are called for.

Resurgence of interest in some types of technology

During the past decade interest in particular types of innovations has come from at least two sources. First, there has developed an interest in the environmental consequences of some of the new technology now used in agriculture. And second, there has been interest in capital intensive innovations, again from the perspective of effects, but in this case effects on the structure of agriculture itself. We will comment on each of these foci of interest, and try to relate them to earlier efforts in diffusion research.

Environmental concerns led to singling out pollution as an effect of current agricultural practices. As a result of the agricultural transformation, specific problems of waste disposal, soil erosion, and water pollution have arisen in the context of agriculture. Government bodies have considered, and in some instances have put in place, a number of remedial actions such as restrictions on the use of pesticides, restrictions on disposal of animal waste, and control of water pollution from agricultural sources (*cf.* the discussion of regulation as a means of control over technology in the chapter by Mazur, later in this volume).

Introducing these policies into agriculture brought to the forefront a number of problems for diffusion research. Of greatest importance, unlike the technology introduced in the past, the changes in agricultural practices which are currently advocated are not justified in terms of farm-firm benefits, but in terms of the public good. The reasons for reducing the use of fertilizer, or prohibiting certain types of cultivation, are not because they will provide the farmer with either short run or even long run benefits, but because they contribute to preventing water supplies from being polluted, or because they have public health implications.

As was the case with the earlier technological innovations, policy makers look upon the issue as one of persuading individual farmers to voluntarily change their farming practices. The reasons for this focus on the individual decision maker are partially ideological, and partially technical. Voluntarism is the preferred ideal in any case. However, even if the political will existed to impose a mandatory program of environmental quality control in agriculture, it does not appear that either the technical or legal resources exist to make such programs work unless there is voluntary compliance on the part of the farm operators.

This emphasis on the individual decision maker approach has led to attempts to rely heavily on past diffusion research to design strategies for bringing about the desired changes in agriculture. However, the new realities highlight the ways in which diffusion research had defined the issue: individual economic benefits and voluntary individual action were assumed, even though these assumptions, and their implications, were rarely scrutinized.

Many changes in rural society have taken place for which the individual economic rationale is not at all clear. Much of the support among farm groups for rural education, for example, complete with extra-curricular activities in sports and music, has extremely weak economic justification. Many home and family life improvements programs have also been widely adopted by farm families, even though they too can frequently not be justified in economic terms.

There are also many regulatory programs in agriculture, including public health requirements in the dairy industry, animal waste disposal regulations, pesticide applicator licensing, and local noxious weed control ordinances. These programs, many of them closely related to the technological transformation of agriculture, have not been studied by sociologists as presenting a legitimate field of adoption research. The typical voluntarist assumptions are probably not applicable, but alternatives have not been explored. While the reasons for this neglect are several, the situation serves to highlight the particularistic nature of the diffusion research tradition.

The diffusion model is almost certainly inappropriate to the issue of gaining acceptance of conservtion and, more generally, environmentally benign production technology (cf. Pampel and van Es, 1977). Reasons for the inappropriateness are perhaps not obvious, but nevertheless convincing. Research on the diffusion of agricultural innovations was based on the assumption of profitability for the farmer. Hybrid corn can serve as a prototype of the kinds of innovations that diffusion research was directed toward. The diffusion model assumes that the farmer, as actor, will respond to an economic incentive if he can be made to realize that the potential is there. Society gains when production increases and prices to the consumer are reduced. Environmentally benign production technology is typically oriented to the realization of societal goals (preservation of soil resources for the future, pure water for recreational and consumptive use, and so on), and the interests of the farmer as producer are at best in second place.

Conceptualization of types of technology has not been directly pursued by researchers concerned with adoption of environmentally benign innovations. Stimuli for research in this area come from *ad hoc* concerns with certain kinds of impacts. Thus soil erosion is a concern, stream pollution is a concern, the impact of pest control practices on other than the target species is a concern, and studies directed to understanding the diffusion of corrective innovations have tended to focus on specific innovations rather than on types. A typology based on the primacy of societal versus individual benefits would seem to be appropriate but efforts to develop such a typology are lacking. The possibility for building directly on earlier efforts to develop typologies does not seem to be great, since the earlier efforts were primarily directed toward solution of technical, measurement problems inherent in utilization of the diffusion model. The underlying issue remains the same, however, understanding the diffusion process for one innovation is not a good basis for generalizing to other innovations. The long-run goal of understanding and prediction remains elusive.

Capital intensive technology has also been singled out for attention from the perspective of impacts. It is not only total capital requirements for adoption, but the magnitude of such requirements relative to the resources of the farmer that are of interest. Early diffusion studies ignore distinctions based on capital requirements, for the most part, because differences in resource position among the North American farmers who constituted the samples were not particularly striking. Export of the diffusion model to third world countries, however, made it apparent that much greater inequality in the resource position of farmers had to be taken into account. The extreme inequality in access to land in many Latin American countries was a direct stimulus to paying greater attention to capital requirements. Not only was a lack of capital a direct inhibitor of adoption of improved technology (see, for example, Havens and Flinn, 1975), but it became apparent that those farmers who could adopt such technology improved their relative economic positions, and existing inequality in resource distribution was exacerbated. It is the latter theme, the structural impact of introducing at least some items of technology, which has come to dominate thinking in this area. The implied shift from a micro- to a macro-level perspective is perhaps most obvious here.

Capital intensiveness is probably not a good basis for sorting agricultural production technology into types. We noted above that it is the magnitude of capital requirements relative to the farmer's resource base that is of interest. Even hybrid corn seed, infinitely divisible, requires a capital outlay because it cannot simply be obtained from last year's harvest. The example given points to the relevance of divisibility, or lack of it, of some innovations, or conversely, the fact that some innovations are not scale neutral. As the scale of farming operation increases, the availability of resources to adopt relatively "costly" innovations tends to increase. Deep well irrigation technology and many mechanical innovations, *e.g.* tractors, are examples of indivisible innovations which tend to involve relatively large capital investments.

The reference to tractors permits one to introduce another characteristic of capital intensive innovations, the fact that capital, especially in the form of machines, often displaces labor. The structural consequences of technology are of interest at both extremes of social structure. Large-scale farmers are disproportionately able to benefit from some items of technology, and small-scale farmers and landless laborers may be displaced by labor-saving technology. Machines, in short, are probably the sub-set of innovations which have the most obvious structural implications and which represent a type of technology worthy of further research (Berardi, 1981).

Implications for future work

The discussion in this paper has been roughly chronological. Diffusion research was traced from an early concern with analyzing the diffusion process for hybrid corn, through a period in which adoption of innovations at a more general level was of concern, to a more recent period in which the consequences of adoption are the central point of interest. Throughout this period of about 40 years there has been some effort directed at specifying types of technology for particular attention. Since early research on diffusion was directed toward finding ways to bring about adoption more quickly, and the main stimulus for more recent research stems from the consequences of adoption, it is understandable that the specification of types of technology has not been a cumulative process. Disaggregating contemporary agricultural production technology in a way which leads to better understanding of the change processes in agriculture is still in its infancy, although a substantial body of literature can be drawn on for guidance.

The diffusion model with its emphasis on antecedents of adoption still has some utility in research and policy formation, especially in those developing countries which mirror most closely in their agricultural structure the features which preceded the U.S. agricultural transformation, namely a large category of farm decision makers with relatively equal access to resources, operating in an agricultural environment with substantial marginal returns to technology.

The most obvious utility of diffusion research in the classic tradition is in particular, local situations, where immediately useful information for purposes of advocacy has high priority. Such highly applied work does not lend itself well to a concern with building a knowledge base, however. Furthermore, there has been a shift in agricultural development policy toward de-emphasizing the distinctiveness of items of technology, and this shift in approach has implications for diffusion research. Advocacy of particular innovations gradually has given way to an emphasis on "packages of practices," which approach stresses the importance of interactions among production inputs. More recently, the favored approach has been via "farming systems," which places even greater stress on interactions. The point of interest in this gradual shift in approach is that the assumed decision-making latitude of the farmer is decreased and shifted from a focus on single innovations to aggregates. For that reason alone, a focus on types of technology as relevant to adoption decisions takes on a quite different meaning.

When one broadens one's perspectives on the diffusion of technology to include the consequences of adoption there would seem to be considerable scope for a sharper focus on types of technology. The bare outlines of two typologies were suggested in the preceding discussion. First, it may be useful to distinguish between technologies which benefit the individual directly and immediately, and those technologies which benefit the larger society most directly. And second, it may be useful to distinguish between technologies which enhance the productivity of land, without causing labor displacement, and those which enhance the productivity of labor and for which the chief aim is labor displacement. Each of these possibilities will be discussed in somewhat greater detail below.

Making a distinction between technologies which benefit society primarily versus those which benefit the agricultural producer in the short run is useful from the perspective of defining researchable problems. The policy issue is how to gain individual acceptance of practices which are not attractive in terms of individual economic gain. Comparative studies on strategies for compensating or penalizing the non-acceptor of societally beneficial innovations are called for. The greatest gain from making the suggested distinction lies in turning away from the fundamentally profit oriented individualistic perspective of the diffusion model to ask the more nearly appropriate research questions. Mazur's discussion of various "levers" for control over technology, later in this volume, is directly relevant here. Similarly, Schnaiberg's distinction between "ideas" and "interests," also in this volume, relates well to the specification of public versus private ends served by technology. The interests of the agricultural producer are central to the adoption of environmentally benign technology and it is quite clear that those interests are at least partially in conflict with societal goals.

A specialized focus on labor-displacing technology is of greatest importance in developing countries for the simple reason that a high proportion of the total labor force is engaged in agriculture in those countries. The issue is also important in some sectors of North American agriculture, but it is of much greater importance where labor is plentiful and access to land is not. North American agriculture has historically enjoyed a plentiful supply of land, and labor has been the scarce resource for the producer. Diffusion studies have paid some attention to "labor-saving" technology (*e.g.* Fliegel and Kivlin, 1966). "Labor saving" has quite obviously benign overtones, and from the perspective of the row-crop, dairy, hog and beef producers of much of the North American farm landscape this is entirely appropriate. Labor displacement, with equally obvious negative overtones, is an issue in North American fruit and vegetable production, and in tobacco, but it is a much more general problem in densely populated countries.

High priority should be given to an historical analysis of the role of labor-displacing technology in the structural transformation of agriculture in the industrialized nations. Very little research has been done on this topic, and, unfortunately, research on the topic does not lend itself well to use of the familiar survey techniques or the preference for immediately useful research results which dominate the rural social sciences. The historical data available to assess the topic are scanty but the topic is important. A structural transformation of agriculture has taken place and neither the process itself, nor the role of technology in it, nor the consequences of that process are well understood (CAST, 1982). By and large it has been assumed that the transition of labor out of agriculture is a natural response to market forces, ultimately benign in its consequences and therefore not worthy of study from a broad, societal perspective. The current, and not yet well defined interest in the "structure of agriculture" may serve to alter our assumptions about what is and is not "natural," but that remains to be seen.

With reference to developing countries, the immediate utility of special attention for labor-displacing technology may lie in the clarification of development policies and procedures. The distinction between labor productivity and land productivity is not always clearly made (*cf.* Shingi, *et al.,* 1981), and this can lead to an inadvertant advocacy of labor-displacing technology which has undesirable societal consequences while doing little to increase food supplies (see Thiesenhusen chapter in this volume). Research on small-scale devices which complement the introduction of new seeds, fertilizer, and pest control practices should have high priority. Finally, some conventional "adoption" research may be useful to assess the acceptability of appropriate technology as it becomes available. But above all, research in this area must focus on the potential for structural consequences when labor-saving technology is under consideration.

While structural change may be "natural," particular structural consequences of technological changes are not necessarily inevitable.

Summary

Our efforts to trace the history of diffusion-adoption research over the last 40 years have permitted us to speculate on two interesting parallels. First, interest in diffusion phenomena grew out of micro-level concerns with farmers' willingness to adopt elements of that universe of technological innovations which ultimately were prominent in the transformation of the industry. As agriculture in the industrialized nations was transformed, the salience of questions about willingness to adopt innovations declined. Instead, new questions about the consequence of widespread adoption of technology gained in priority. And second, while the individual actor focus of the discipline of sociology served the early diffusion researchers well, export of the research tradition to the developing world raised questions about social structure which had not been considered earlier. Shifts in theoretical preferences in the discipline as a whole ran more or less parallel to shifts in the empirical perspectives of diffusion researchers. The outcome of that change process has been a gradual movement from an individual actor-in-situation to a structural and societal perspective in problem formation.

The diffusion-adoption research tradition remains viable after 40 years. Different research questions are being asked, and modifications in conceptual approaches and the techniques of doing research are called for, but the relationship between technology and society continues to be problematic.

We have attempted to sketch out two directions for contemporary diffusion-adoption research with the objective of enhancing the continued viability of the research tradition. The first of these new directions calls for a distinction between the public versus private ends to be served by technology and has particular relevance for the environmental consequences of agricultural production technology. We take it as given that environmental protection and resource conservation are societal goals, and we infer that the historic individual actor approach of the classic diffusion model is inappropriate for research in this area. Research directed toward the efficacy of group actions is called for. The second new direction for research singles out a different type of technology, in this case mechanical, or more broadly, labor-saving technology. The problem focus is again the societal impact of such technology, but micro-level studies are needed in the developing countries if the structural consequences of labor-saving technology are to be understood.

Retrospective studies, from a broader, societal perspective would also be useful in the industrial nations, where the structure of agriculture has already been transformed but the transformation process itself is not well understood.

New paradigms are clearly appropriate, but the classic diffusion model has a role to play in the research of the future as well.

References

Berardi, G.M. 1981. "Socio-economic consequences of agricultural mechanization in the United States: Needed direction for mechanization research." *Rural Sociology* 46 (Fall): 483-503.

Council for Agricultural Science and Technology (CAST). 1982. *Agricultural Mechanization: Physical and Societal Effects, and Implications for Policy Development.* Ames, IA: CAST, forthcoming.

Fliegel, Frederick C., and Joseph E. Kivlin. 1966. "Attributes of innovations as factors in diffusion." *American Journal of Sociology* 72 (November): 235-248.

Flinn, William L. 1982. "Rural sociology: Prospects and dilemmas in the 1980s." *Rural Sociology* 47 (Spring): 1-16.

Galjart, Benno. 1971. "Rural development and sociological concepts: A critique." *Rural Sociology* 36 (March): 31-41.

Goss, Kevin F. 1979. "Consequences of diffusion of innovations." *Rural Sociology* 44 (Winter): 754-772.

Gross, Neal. 1949. "The differential characteristics of acceptors and nonacceptors of an approved agricultural technological practice." *Rural Sociology* 14 (June): 148-158.

Gross, Neal and Marvin J. Taves. 1952. "Characteristics associated with acceptance of recommended farm practices." *Rural Sociology* 11 (December): 321-327.

Havens, A. Eugene and William L. Flinn. 1975. "Green revolution technology and community development: The limits of action programs." *Economic Development and Cultural Change* 23 (April): 469-481.

Kivlin, Joseph E. 1960. "Characteristics of Farm Practices Associated with Rate of Adoption." Unpublished Ph.D. dissertation, Pennsylvania State University, University Park, PA.

Lionberger, Herbert. 1960. *Adoption of New Ideas and Practices.* Ames, IA: Iowa State University Press.

Pampel, Fred, Jr., and J.C. van Es. 1977. "Environmental quality and issues of adoption research." *Rural Sociology* 42 (Spring): 57-71.

Rogers, Everett M. 1962. *Diffusion of Innovations.* New York: Free Press.

Rogers, Everett M., with F. Floyd Shoemaker. 1971. *Communication of Innovations: A Cross-Cultural Approach.* New York: Free Press.

Ryan, Bryce. 1948. "A study in technological diffusion." *Rural Sociology* 13 (September): 273-285.

Ryan, Bryce and Neal Gross. 1943. "The diffusion of hybrid seed corn." *Rural Sociology* 8 (March): 15-24.

_____. 1950. *Acceptance and Diffusion of Hybrid Corn Seed in Two Iowa Communities.* Ames, IA: Agricultural Experiment Station, Iowa State College, Research Bulletin 372.

Shingi, Prakash M., Frederick C. Fliegel, and Joseph E. Kivlin. 1981. "Agricultural technology and the issue of unequal distribution of rewards: An Indian case study." *Rural Sociology* 46 (Fall): 430-445.

Wilkening, Eugene A. 1954. "Change in farm technology as related to familism, family decision-making, and family integration." *American Sociological Review* 19 (February): 29-37.

Wilkening, Eugene A., Joan Tully, and Hartley Presser. 1962. "Communication and acceptance of recommended farm practices among dairy farmers of Northern Victoria." *Rural Sociology* 27 (June): 116-197.

Chapter 2

Public Protests Against Technological Innovations

Allan Mazur

Science and technology are often regarded as proceeding under their own inertia, largely outside of the control of human actors. In reality, there are three effective levers that control the pace and direction of research and development. Two of these — funding and regulation — may be adjusted by the federal government and industry as a part of normal policy making. My concern here is with the third level, public protest, which has affected nuclear power, fluoridation, the SST, recombinant DNA research, and many other technological innovations (Nelkin, 1979). Unlike funding or regulation, public protest cannot be turned on or off by a policy decision.

Why do some technologies become the focus of widespread public protest while others, no less hazardous, are tolerated? If nuclear power plants are challenged because of the risks to health, and the supersonic transport was opposed for its possibly deleterious effects on the ozone layer of the atmosphere, then why aren't similar protests raised against fossil-fuel power plants which cause an appreciable mortality rate and may change the earth's climate through a buildup of carbon dioxide? Where is the public protest against the automobile, which kills over 40,000 Americans per year and keeps us dependent on foreign oil? Why is there no citizen opposition to the proliferation of extraneous x-rays, drugs, and

This chapter is reprinted, with some changes and updating, from Chapters 7 and 8 of *The Dynamics of Technical Controversy*, by Allan Mazur (Washington, D.C.: Communications Press, 1981). I appreciate the contribution of Peter Leahy and the advice of William Gamson, Louis Kriesberg, Robert Mitchell, and Gene Summers.

29

surgery promoted by the medical and dental industry, while major court battles are waged to stop the Tellico Dam in order to protect a small fish, the snail darter, one of hundreds of species listed, for administrative purposes, as endangered?

These questions are so complex as to be intractable. The most I can attempt is a helpful analysis, breaking the problem into simpler components. We can view the growth of protest movements as a process which takes place in three successive steps, recognizing that the discreteness of these steps exists more in the mind of the analyst than in the real world.

In the first step, a warning against a technology is brought to public attention and is usually noted in the public literature (i.e., the mass media, a book, the *Congressional Record,* etc.). Reasons are given for opposing the technology — beyond simply monetary costs — and some corrective action may be suggested. Responses from the promoters of the technology, refuting the critics, also appear in the public literature at this time.

The protest may go no further, or it may move to a second step in which a small number of people, organized into at most a few action groups, attempt to stop or modify the technology, perhaps through legal procedures such as a lawsuit, or through lobbying, protest demonstrations, or letter writing campaigns. The essential feature of this second step is that the activity is bounded, or limited, to a small number of well defined groups, and the interest of the mass media is restrained. If the number of protesting units multiplies quickly, and they become joined into a more or less organized network, and many previously uninvolved people are recruited into the protest, then we enter the third step, the regular mass movement so enjoyed by the public media. These steps, and intermediate links between them to be discussed below, are illustrated in Figure 2.1, which provides a convenient structure for this discussion. In the next three sections I will describe each step in more detail.

Step 1. A Warning is Brought to Public Attention

Many new technologies are initially challenged by the promoters themselves, or their associates, and are settled before we on the outside ever hear of them. Inhouse challenges are most likely to occur when there is a preexisting cleavage within the community of professionals involved with the technology. It is unreasonable to expect a proud promoter to be a strenuous critic of his own invention, but if the promoter belongs to one of two competing divisions of a company, or to one of several competing companies, then challenges are likely to come from the competition. In technologies heavily dependent on federal money, government agencies and congressional committees which oversee these funds become integral parts of the involved community, and there is preexisting cleavage along the lines of party or other voting blocs, which encourages challenges to the proposed technology. Several space and weapons systems have been promoted and then stopped by such opposing interests without ever becoming public issues of any import, for example, the nuclear airplane (Lambright, 1967).

FIGURE 2.1

Three Steps in the Growth of Protest

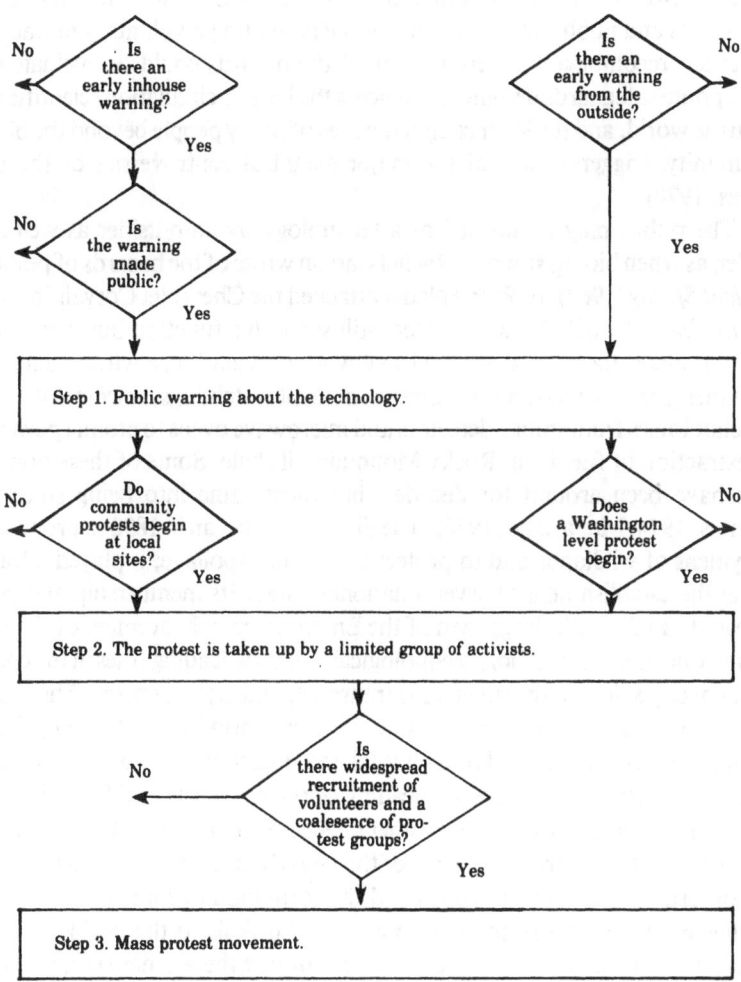

Within the esoteric world of molecular biology there were methods developed in 1973 which allowed the splicing together of DNA (the genetic material) from different species of organisms. This recombined DNA, when inserted into a bacterium, would replicate as if it were the normal genetic complement, though in fact it could be a completely new hybrid. Some of the early recombinant DNA workers became concerned that these new life forms could be extremely toxic or tumor forming, and that they might excape from normal laboratory confinement, spreading to the population. A clique of these prominent researchers formed themselves into a committee under the auspices of the National Academy of Sciences and published a letter in *Science* requesting a voluntary moratorium on certain recombinant experiments until the hazards could be evaluated and appropriate safeguards designed. *Science* is the largest-circulation scientific journal in the world, and the letter caught the eyes of many people beyond the biology community, triggering one of the major technical controversies of the 1970s (Ames, 1978).

The public may be warned of a technology by an outsider as well as an insider, as when biologist-writer Rachel Carson wrote of the hazards of pesticides in *Silent Spring* (1962), or Ralph Nader attacked the Chevrolet Corvair in *Unsafe at Any Speed* (1965). Isolated writers still serve this function but increasingly, since the 1960s, there has developed a network of scientific, environmental, and consumer groups which have become permanent watchdogs, warning of a range of technologies from smoke detectors and microwave ovens to atomic power and the extraction of fuel from Rocky Mountain oil shale. Some of these organizations have been around for decades, but most came into being since 1960 (Nichols, 1974; McFarland, 1976). The Sierra Club began in 1892 to promote the enjoyment of wildlands and to protect them from spoilage; it played a leading role in the establishment of several national parks. Its membership and power increased rapidly with the growth of the Environmental Movement of the 1960s and its concerns became more technological as it took leading roles in the controversies over pesticides, the supersonic transport, and atomic energy. The Federation of American Scientists began shortly after World War II to lobby for the control of nuclear arms, and though it fell into inactivity soon afterward, it was revitalized in 1969-70 to become a leading opponent of the anti-ballistic missile (ABM) and has remained active on other science-related issues. Most of the political causes of the remarkable decade of the 1960s had some impact on the changing concerns of these older groups and the formation of numerous new ones, whether it was the desire to apply our scientific skills to the problems of the ghetto, or the revolt against military research during the Vietnam era, the rise of consumerism, or the wreck of the environment by technology. These groups and associated individuals — lawyers, scientists, writers, environmentalists — have a large supporting membership, excellent access to the mass media, and a broad range of publications which they put out themselves. They constitute a potent political force on the Washington scene, in the courts, Congress, and the Executive Branch. Many alumni of this network have served in prominent government positions, and these personal links provide access to the federal policy process. However, their resources are limited compared to their corporate adversaries, and they seem particularly attracted to fashionable causes, ignoring many that are not.

Sometimes the initial challenge to a technology comes from the grass roots, with opposition arising spontaneously in communities near a site where the technology is to be implemented. The first public opposition to fluoridation was raised in Stevens Point, Wisconsin, led by an elderly man long active in local politics, who didn't want chemicals dumped into the town water supply (McNeil, 1957). The initial opposition to nuclear power plants, back in the 1950s and early 1960s, was raised by local groups over plants proposed for their areas. Several new airports, or proposed enlargements or existing airports, have been opposed by local citizens groups (Milch, 1979; Nelkin, 1974). Recent opposition to very high-voltage electric-transmission lines was generated primarily by farmers whose lands would be crossed by the lines (Gerlach, 1978; Casper and Wellstone, 1981). These local challenges are usually an attempt to protect one's personal interests from infringement by external agencies, particularly the state or large corporations. Initially the technology is seen as unacceptable because it is being sited here rather than somewhere else, though as the protest evolves, the technology may come to be opposed on its own, no matter where it is located.

There are, of course, federal agencies specifically concerned with the identification and assessment of hazards to health and environment. Early in the century, the Food and Drug Administration used to verify the safety of various substances by feeding them to volunteers, a practice now discontinued, though apparently not for lack of willing participants. This method was not quite as foolhardy as it may appear at first, becuase the drugs of that time were relatively ineffective, for good or ill. However with rapid strides in pharmacology's ability to do us in, better evaluative techniques were called for.

During the 1960s and 1970s there were moves to increase regulatory activities of the federal government, with the attendant creation of new agencies. Among the most important are the Environmental Protection Agency, the Consumer Product Safety Commission, the Occupational Safety and Health Administration, and the Congress' Office of Technology Assessment. By most accounts, the tasks given these agencies overwhelm the resources which they have available for the work, particularly since the budget cuts of the Reagan administration.

All told, challenges and warnings about certain technologies reach public attention from a variety of sources both inside and outside of the community of promoters, and from the grass roots of the nation as well as Washington. This does not appear to be a particularly efficient screening system since many hazardous technologies are ignored until damage is done, and excessive attention may be focused on a few technologies which are not inordinately risky.

However, there can be no doubt that real dangers are discovered and recorded in the public literature for everyone to see, if they would look.

Step 2. The Warning is Taken Up as a Bounded Protest

Once a warning is made public, it may or may not be taken up by a group that is willing to act as an adversary to the promoters of the technology. Technological products for which warnings have been raised, but public protests were never strenuously pursued, include birth control pills and numerous other drugs, color television, microwave ovens, weather modification, medical and dental x-rays, numerous food additives, aluminum cookware, wood burning stoves, gas furnaces, electric wiring in the home, chlorination of water, swine flu vaccine, several surgical procedures, smoke detectors, hybrid seeds, the internal combustion engine, computer data banks, dams, and so on. In most cases, the warnings were little noticed and barely remembered.

In cases where the warning is taken up, the incipient protest tends to be bounded in the sense that the number of protesters is relatively small, only a few easily identifiable groups are involved, and they are limited to a few geographic sites. Usually the attention of the mass media is limited at this point, and the general public is uninformed, perhaps unaware of the controversy.

The early protest will be played out either at the local level, in a community where the technology is to be sited, or at the federal level, which usually means in Washington. The particular level, local or federal, is determined in large part by the manner in which the technology is to be deployed. If it is to be deployed in designated sites around the country, as in the cases of fluoridation, nuclear power plants, or electric transmission lines, then the earliest protests occur near proposed sites (Coleman, 1957). If the technology is not associated with particular sites, either because it is spread uniformly throughout the population, or because it is national in character, then the protest will focus on Washington. Thus the controversy over the airbag, which was to be required in all automobiles as a crash safety device, took place wholly in Washington because there were no salient locales as alternatives. Since the ABM was a national-level weapons system, its early protest also focused on Washington, though later, when certain cities were designated as proposed missile sites, these became foci for local disputes.

Step 3. Mass Movement

In 1977 farmers in both Minnesota and upstate New York began protests against very high-voltage electric transmission lines which were to run across their lands. These were new types of lines, of higher voltage than had been used previously.

Since many farmers resisted initial utility attempts to obtain easements through their properties, the utilities exercised their legal rights to obtain these rights-of-way forcibly. A major issue in the controversy became the utilities' violation of individual rights to land. A second issue was the hazard from electromagnetic fields generated by the lines, a hazard which the protesters claimed to be greater than the utilities recognized. This technical argument was developed by two medical researchers from upstate New York who brought national attention to the lines, if only briefly, by their appearance on the televison program, "60 Minutes."

There was a good deal of popular support for the farmers' cause in both states, and the power lines became an issue in gubernatorial politics, particularly after some incidents of sabotage against construction work which were dramatized in a made-for-television movie that was shown nationally. In spite of the efforts of the farmers and their allies, the lines in both states were completed in 1979, and the protests appear to be over, though it is too soon to be sure that they will not reemerge. Throughout the two-year controversy, from the initial warning to the completion of the lines, the protest remained bounded, being limited to fairly well-defined groups at the two sites; there was not even much communication between the Minnesota protesters and those in New York.

In many ways the fluoridation controversy is similar to the power line controversy. Both began in small communities which had been chosen as early sites for a new technology. In both cases, vocal citizens feared that the technology was hazardous, and perhaps more important than the hazard, they felt that the technology was being forced down their throats in spite of their objections. These activists became the leaders of community revolts. At this point the stories diverge. In two years of transmission line controversy, only two communities protested, partly because there were only a few other sites where very high-voltage lines were being put up. In contrast, several communities considered adopting fluoridation within the first two years of that controversy, so there were many sites available for protest. Furthermore, several of the early protest groups managed to reject fluoridation, while both of the transmission line protests failed. The two transmission line protests remained bounded and virtually isolated from each other. In contrast, the fluoridation protest quickly burst its bounds; the number of community protests increased rapidly, drawing in numerous previously uninvolved people during the course of referendum campaigns. Formal and informal links grew among the local protest groups, lending support and exchanging information. Within a very short while, the opposition to fluoridation had developed into a mass protest movement of national proportions.

Why did the transmission line protest remain bounded and eventually fail, while the fluoridation protest mushroomed? A strategy that has succeeded in other protest movements, not only fluoridation but nuclear power and the ABM as well, has been to promote local protest groups, using them as building blocks to form a national coalition. What would have happened if an organizer had encouraged communication between the Minnesota and New York groups, forming an alliance of mutual support? The medical researchers who had effectively raised the hazard issue in New York, bringing it to a national television audience, could have lent weight to the Minnesota protest. The two states, now linked together, might have promoted protests in other locales, such as California, where very high-voltage power lines were being dicussed, and then brought these groups into the coalition. Perhaps the building momentum would have produced a national protest.

A coalition of local protest groups is one of two pillars of most successful national movements, the other being a strong effort in Washington to influence policy through the federal courts, Congress, and the agencies and departments of the Executive Branch. The fluoridation controversy was unusual in that the opposition worked mostly at the community level, without an effective Washington lobby. In most cases coalitions have tied the community protest groups to Washington, the usual belief being that skirmishes in the field necessarily have a limited effect, and that the death blow must be struck in the capital at the center of power. The giant nuclear power controversy may be taken as a paradigm of this philosophy.

The technical protests which grow to the size of a national movement have a number of features in common with large nontechnical movements (Gerlach and Hine, 1970; Obershall, 1973; Kriesberg; 1973, 1981), so it is possible to draw a fairly robust picture of a generalized movement. With a few variations, the controversies over nuclear power, fluoridation, and the ABM all fit this pattern: local opposition groups emerge with officers, by-laws, and dues; and these become linked by various personal and organizational ties into a national coalition which supports campaign and lobbying efforts, both locally and in Washington. (Dispersed local protests may precede the Washington effort, as in the nuclear power controversy, or follow the initial Washington action, as in the ABM case.) A few individuals on each side of the controversy emerge as nationally known spokespersons, and opposing, hostile camps become clearly identified, one the establishment side supporting the technology, the other the challenge side made up primarily of voluntary organizations. Experts buttress each side's position with technical arguments which may seem contradictory to the layman. New members are recruited to the challenge side, often from among friends and acquaintances of those already involved. Frequently new members join as a bloc when organizations to which they belong merge with the coalition of protest groups. Public demonstrations help draw the attention of the mass media to the controversy, and the carnival-like atmosphere of these demonstrations may attract protesters whose interests are transient. Movement participants develop a shared outlook which emphasizes the hazards of the technology and their confrontation with the establishment. This outlook is frequently expressed in a stereotyped rhetoric which fits into a pattern of attack and rebuttal with the equally stereotyped rhetoric of the promoters of the technology.

The evolution of a protest into a mass movement is critical for public partic-ipation, because it is only at this stage that opportunities are available for lots of people to add their efforts together to produce a desired change. Most of us would have no effect working alone, nor would we receive any social support to sustain our actions. I may write letters to editors and congressmen about any of my favorite problems, but these will be straws in the wind if there are no other let-ter writers, and if I have no better ways to mobilize opinion and to apply pres-sure. The average person cannot effectively lobby Congress, or obtain television coverage for a favored cause, or make speeches on a campus tour. But if he can-not start a movement, he can join one that already exists and that provides easy channels for protest, such as letter-writing campaigns, marches, meetings, news-letters, rallies, and the like. The options usually available to someone who wants to express his concerns are limited: either waste one's efforts on solitary protests with little chance of success, or join a currently-running protest movement and pool resources with other sympathetic souls. Joining a preexisting movement also offers the emotional reward of marching with one's fellows in a just cause, a feeling denied the solitary activist.

Rise and Fall of Controversy

Once protests have become mass movements, their rise and fall is determined, to a great extent, by rising and falling levels of public interest in larger issues that are relevant to the technologies. Morrison provides a detailed development of this point in his analysis of the approprite technology movement, Chapter 11 in this volume. For example, the movement against nuclear power gained great strength during the late 1960s when the American public became massively concerned with the issues of environment and pollution. It is during such periods that pro-test leaders are best able to raise funds, recruit members, find allied organiza-tions, and have access to the mass media (McCarthy and Zald, 1977). In short, these are the times when social resources become widely available to support movement activities.

By developing quantitative indicators of movement activity, I will show that fluctuations of opposition to nuclear power and fluoridation follow an orderly sequence in each controversy. First, challenges from activists increase during periods of national concern with a major issue closely related to the controversy. This national concern provides a supportive milieu in which the activists can obtain new members, money, media attention, etc. As activism increases, spurred on by the new availability of societal resources, the attention of the mass media turns increasingly to the protest, partly to report the activities of the pro-testers and partly to carry protechnology propaganda which is designed to refute the challengers. The nation then becomes a spectator to the dispute through the press and television (MacKuen and Coombs, 1981). The increasing prominence of the controversy in the media is followed by increasing opposition to the technology within the wider public, partly because previously inattentive people have now become concerned. Mass media attention soon drops off, perhaps because of a saturation effect, or more likely because of competing issues vying for attention, and opposition among the wider public drops off as a result. If the activists raise their level of protest again, the mass media again increases their coverage, and opposition again increases among the wider public.

This model depends heavily on the distinction between activists and the general public. The activists express their opposition to a technology like fluoridation in terms of larger national issues like socialism and individual rights. During periods of high national concern with these larger issues, the activists' latent opposition is spurred, aided by sympathetic supports. The wider public is essentially passive, but as they become aware of the controversy through the media, they become increasingly skeptical of the technology, increasing their opposition responses in opinion polls and referenda.

Coverage that mass media periodicals give to the technology, and to the dispute around it, may be measured by the yearly number of articles indexed under each technology in *Readers' Guide to Periodical Literature*. The *Readers' Guide* has an obvious shortcoming, ignoring television and the daily newspapers, but it does give a good view of the interests of a wide range of magazines and journals in the United States.

The opinion of the general public toward a technology may be measured by repeated opinion polls. It is rare to find the ideal situation of the same polling organization asking the same question, frequently, at fixed time intervals, using samples drawn in a similar manner each time. However, useable time series are available for nuclear power plant and fluoridation issues.

The activity of protesters is especially difficult to measure. It is desirable to have an indicator that would be applicable across all controversies, but none is available. For the early years of the nuclear power controversy, it is reasonable to count the number of nuclear power plants, or proposed plant sites, where citizen groups had recently intervened as a measure of activism (Mazur, 1975). However by the early 1970s nearly all such sites were opposed, so the indicator becomes saturated, unable to measure further increases in protest. Also, important aspects of protest, such as lobbying in Washington, may occur independent of activity at local sites, particularly as the controversy has matured. The yearly number of defeats for fluoridation in community referenda is a useful measure of leadership activity since the organization of a referendum usually indicates organized opposition, particularly when the proposal is defeated (Crain, et al., 1969). Since so many communities were available as fluoridation sites, this indicator does not have the saturation problem of the nuclear plant sites; also, the bulk of antifluoridation activity went on at the community level with little Washington activity, unlike the nuclear power case.

The indicator trends for each technology are shown in Figures 2.2 and 2.3. Taking each case in turn, a sympathetic eye will see, first, that *rises in media coverage follow rises in protest activity,* and second, that *rises in public opposition* (as measured in opinion polls) *follow rises in media coverage.*

The first communities to fluoridate encountered little opposition, but then a raucous conflict developed at Stevens Point, Wisconsin, an area of concerted activity by dentist-proponents, and fluoridation was defeated by town referendum in 1950. Since that time fluoridation has spread, though each year some communities still reject it in referenda. I have taken yearly number of referendum defeats as my measure of activity among antifluoridation leaders, but this indicator has built-in time lag since a year or more may pass between the initiation of a referendum and the vote. Comparison of trends in the early years of the controversy is further complicated because the *Readers' Guide* was published biannually until 1965, so the dating of yearly peaks of mass media coverage is relatively imprecise. Several local groups of activists began referendum campaigns shortly after the Stevens Point defeat of 1950, and many defeated fluoridation by 1952. These disputes are discussed in the periodical output which peaks in about 1952 (Figure 2.2). Public opinion polls show increasing opposition between 1952 and 1953, following the article peak, and then decline. We cannot follow opinion change between 1956 and 1959 because the question changed. (Complete sources and details appear in Mazur, 1981:117.) There is a sharp rise in referendum defeats in 1964 followed, in 1965, by a new peak in mass media coverage and also by a relative high in opposition in the opinion polls. All indicators again decline until 1970 when there is another rise in referendum defeats accompanied by a rise in periodical articles. Unfortunately no opinion poll was taken in 1971, but the 1972 poll shows an increase in opposition from 1969.

The first citizen intervention, against Detroit's Enrico Fermi nuclear power plant, occurred in 1956 but is isolated in time. There was a great deal of mass media coverage of atomic power at that time, almost all of it positive, particularly the discussions of President Eisenhower's "Atoms for Peace" program which encouraged the development of civilian nuclear reactors. The opponents were barely noticed. The quantitative indicators of controversy in Figure 2.3 begin in 1960, shortly before a small cluster of plant interventions in the early 1960s which did attract mass media attention, particularly the dispute at Bodega, California. No public-opinion trend data are available prior to 1964, so it is impossible to determine if there was a rise in opposition in response to the 1964 peak of mass media coverage, as hypothesized. We can see, however, that opposition in the polls diminished as the output of periodical articles waned. There was little protest in the mid-1960s, but in 1968 local groups of activists began to intervene against most nuclear power plants that had been proposed. Periodical articles increased to cover these disputes and to carry propaganda put out by both sides. Opinion polls, now appearing yearly, show a peak of opposition among the wider public in 1970 following the peak of mass media coverage in 1969. It was

FIGURE 2.2

The Fluoridation Controversy

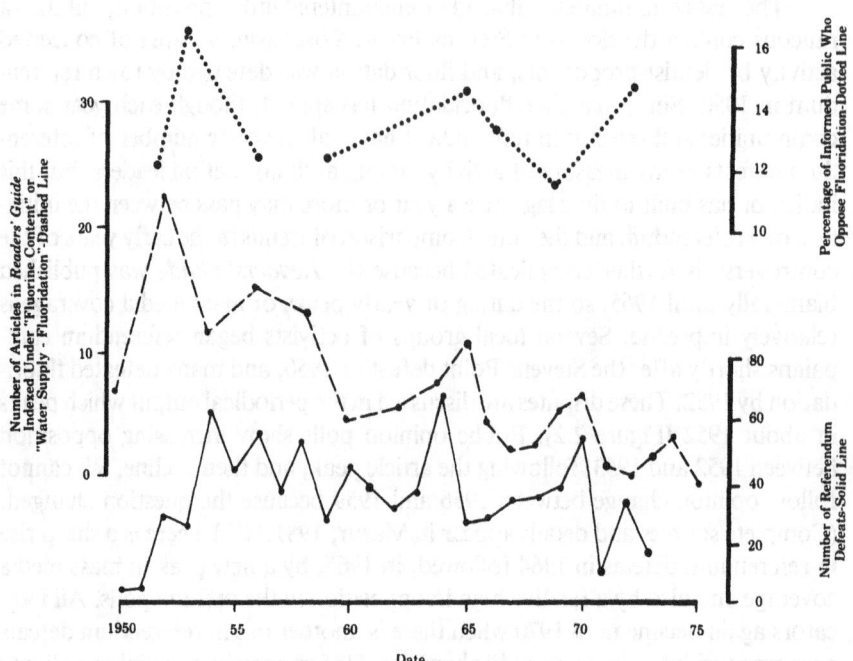

FIGURE 2.3

The Nuclear Power Plant Controversy

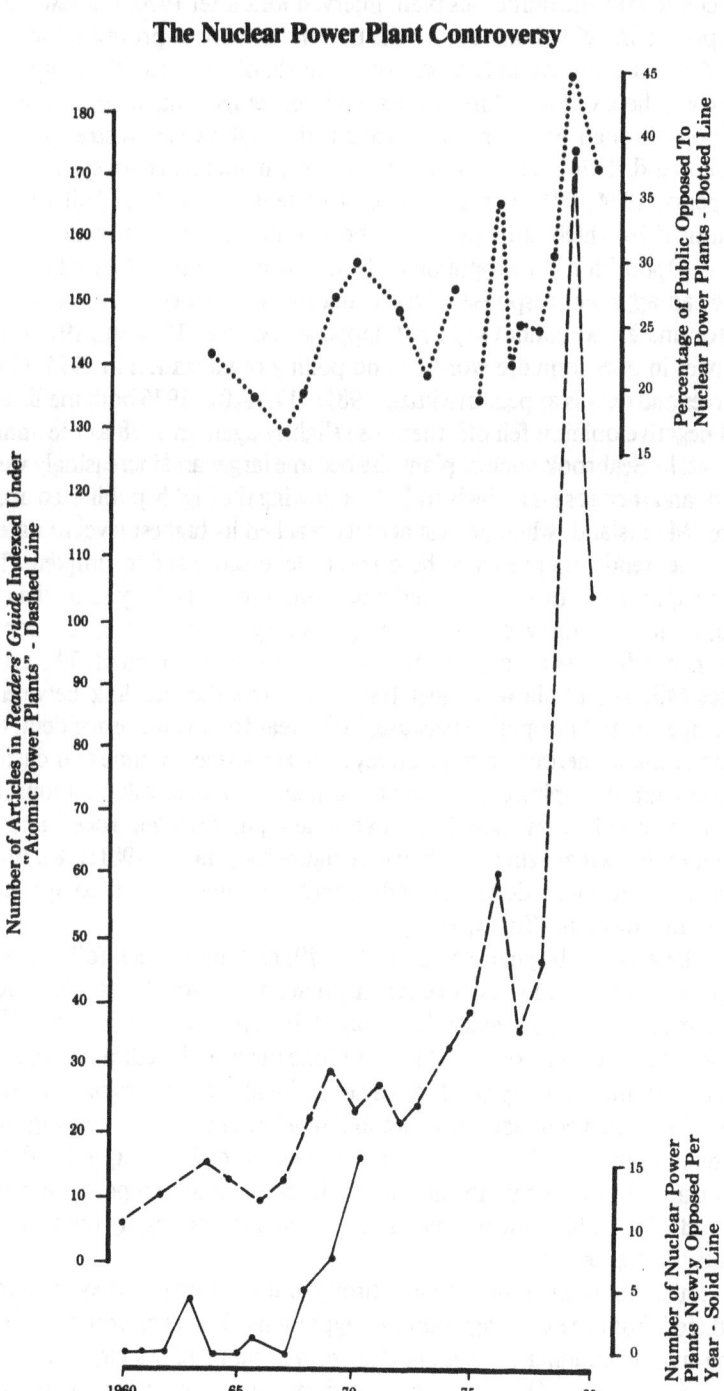

difficult to tally the numerous plant interventions after 1970, but few plants were unopposed in 1971. It appears to me that the activity of protest leaders declined in 1972, but objective indicators are not available to verify that impression. In any case, there was a decline in mass media coverage, and in opposition on opinion polls through 1973. The antinuclear activists showed new strength after 1973 (McFarland, 1976). This is reflected in rising mass media coverage throughout the period 1974-1976, with a great deal of attention paid to a California referendum in 1976 which attempted to impose a moratorium on nuclear power but failed. Opposition in the opinion polls rose from 1973 to 1974, and from 1975 to 1976, an apparent response to the rising mass media coverage of that period. Unfortunately we cannot say what happened between 1974 and 1975 because of changes in questionnaire wording and polling organization in 1975. (Complete sources and details appear in Mazur, 1981: 117.) After 1976 both media coverage and negative opinion fell off, then rose slightly again in 1978 as the annual protests at the Seabrook nuclear plant site became larger and increasingly well publicized, and then rose massively in 1979 following the highly publicized accident at Three Mile Island, when protest activity reached its highest level to date.

The trend data presented here are crude, erratic, and incomplete. To see in them support for my hypothesized links from protest activity to media coverage to opposition in the wider public, requires a sympathetic eye, and others may reasonably disagree because of the large errors of measurement. The accident at Three Mile Island allows a finer test of the hypothesized link between media coverage and public opinion because, in the year following the accident, the Harris poll took numerous opinion surveys, closely spaced in time (Mitchell, 1980). This fine-grained opinion trend can be compared to weekly fluctuations in coverage of the accident on television network news, in *The New York Times,* and in the major news magazines, as shown in Figure 2.4 (Mazur, 1981:118). In order to remove erratic fluctuations, the media trends have been smoothed by the method of running medians (Tukey, 1977).

The accident began on March 28, 1979, and in the week following, almost 40 percent of television-network evening news was devoted to it. Coverage in the news magazines was necessarily delayed by one week, but both *Time* and *Newsweek* ran cover stories in April. By June the story had disappeared from the news magazines and appeared in only occasional short pieces on television until about October, when there was a second, much smaller, rise in coverage to report the final work of the Kemeny Commission which had been appointed by President Carter to investigate the accident. The commission's report was released at the end of October, but the media had been anticipating it, reporting related events throughout October.

Fluctuations in public opinion throughout 1979 are precisely what would be predicted from the coverage-opinion hypothesis. The proportion of the public opposing the building of more nuclear power plants rose sharply after the first burst of coverage. This is hardly surprising since the specter of the accident

FIGURE 2.4

Public Opinion toward Nuclear Power Plants, and Media Coverage of the Three Mile Island Accident

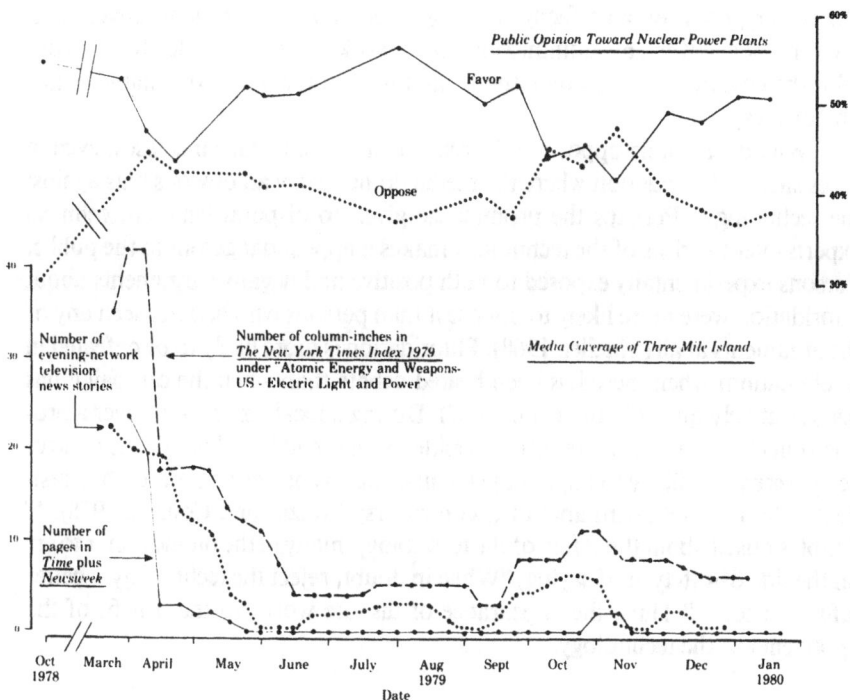

would be expected to increase opposition regardless of any independent effect of the quantity of coverage. However, one would not expect, on this basis alone, that support for nuclear power would rebound within two months, as soon as the media coverage had fallen away, yet that is what happened, in accord with the coverage-opinion hypothesis. Furthermore, a clear, short-term increase in public opposition appeared again during October and November (with rebound by December), coinciding perfectly with the secondary peak of media coverage at the time of the Kemeny Commission's final work. This is completely in accord with the coverage-opinion hypothesis, and is not otherwise explainable in any obvious way.

Why does public opposition increase as media coverage increases, even in cases such as fluoridation where the media do not carry an obvious bias against the technology? Perhaps the prominence given to disputes between technical experts over the risks of the technology makes it appear dangerous to the public. Persons experimentally exposed to both positive and negative arguments about fluoridation were more likely to oppose it than persons who had not seen any of the arguments at all (Mueller, 1968). Fluoridation is more likely to be defeated in a referendum when there has been heated debate than when the campaign has been relatively quiet (Crain, et al., 1969). During a local controversy over a proposed nuclear waste storage facility, residents who had heard about the controversy were more likely to oppose the facility, and to consider it unsafe, than residents who had not heard about the controversy (Mazur and Conant, 1978). If doubt is raised about the safety of the technology, many in the public prefer to err on the side of safety, as if saying, "When in doubt, reject the technology – better safe than sorry." Thus, the appearance of dispute works to the benefit of the opponents of the technology.

Large National Issues

If activists' opposition to a technology such as the nuclear power plant reflects concern with some larger national issue such as the degradation of the environment, then a period of rising national interest in that larger issue might lead to a rise in opposition to the particular technology. The larger issue may itself have the form of a mass movement, for example the Environmental Movement, and then we would have the fusion of two movements, one large and one small, each supporting the other in publicity, enthusiasm, membership, fund-raising, organization and communication, and strategy.

Three large issues appear to account for the major fluctuations in opposition to nuclear power plants. These are the atomic bomb-testing and fallout concerns of the early 1960s, the pollution and environmental concerns of 1969-1972, and the "energy crisis" of 1973 and beyond.

In order to provide an indication of periods of national concern with these

FIGURE 2.5

What is the Most Important Problem Facing the Country Today?

Issues Related to Nuclear Power

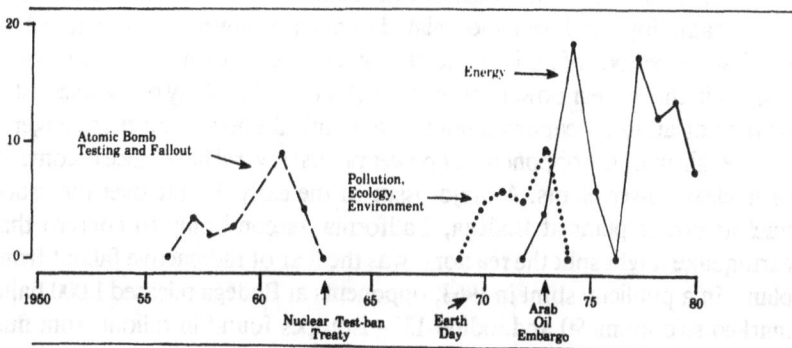

Issues Related to Fluoridation

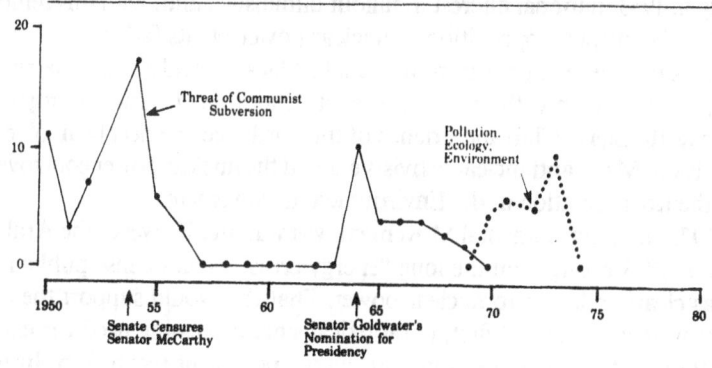

Percentage of Respondents Specifying this Problem

issues and others which will be considered shortly, it is convenient to use responses to the Gallup poll question, which has been asked in most years since 1950, "What is the most important problem facing the country today?" The proportion of the population listing a particular problem in a given year is taken as a measure of national concern with that problem. These trends appear in Figure 2.5 where issues pertinent to each of the controversies are grouped together.

Examining the three issues related to nuclear power plants (Figure 2.5), we see that the periods of major concern coincide roughly with the periods of rising activity in the nuclear power controversy (Figure 2.3). Many of the scientists who warned of atomic weapons fallout were political liberals and thus congenial to the liberalism of the opponents of power plants; several have since become critics of nuclear power plants. A major issue of the early dispute over the proposed nuclear power plant at Bodega, California—second only to concern that an earthquake might split the reactor—was the fear of radioactive fallout from the plant. In a publicity stunt in 1963, opponents at Bodega released 1,000 balloons marked strontium-90 and iodine-131—isotopes found in fallout from nuclear weapons. After the United States and the Soviet Union signed the nuclear test-ban treaty in 1963, national interest in fallout diminished and, with this removal of support, the incipient opposition to nuclear power plants faded.

The resurgence of opposition in the late 1960s coincides with the rise of massive public concern with the environment. Major environmental organizations such as the Sierra Club and Friends of the Earth became deeply involved in the opposition. Many antinuclear activists entered the nuclear power controversy through their participation in the Environmental Movement.

By 1973 the Environmental Movement was waning. However the Arab oil embargo of 1973 brought on the long "energy crisis" as an intense public issue which was clearly relevant to nuclear power. That this would support the antinuclear movement seems, at first, paradoxical since a pressing need for energy might be thought to quell opposition to an energy-producing technology. Instead the increased national concern with energy, like earlier concerns with nuclear fallout and the environment, gave new focus and importance, and a renewed audience, to the movement against nuclear power plants. A new group of activists joined the antinuclear camp, many of them younger and more closely associated with the peace movement than those who were already there. Antinuclear leaders began to express their opposition in the ideological context of energy conservation and the need for new, safer forms of energy such as solar power.

We may visualize the nuclear power controversy as a surfer riding successive waves which are large national issues—first the nuclear fallout wave, then the environmental wave, and then the energy crisis wave. As each wave diminishes, the technical controversy falls unless it can catch another wave. Each wave, to be suitable, must have clear relevance to the technical issue, and it must be politically compatible, or at least not incompatible, with the leadership of the smaller movement.

It is easy, of course, to look back over the course of controversy and "fit" its peaks with various larger issues which were prominent at the time, since there is no shortage of political issues. To avoid the worst excesses of post hoc theorizing, I have set some methodological rules for the identification of these larger pertinent issues. First, I consider only those issues prominent enough to be registered on the Gallup poll of "most important problem," which specifies only five to ten issues for a given sampling. Second, I require an obvious connection between the national issue and the technical controversy. Thus, an energy crisis is obviously related to nuclear power but not to fluoridation. Third, I require political compatibility between the technical controversy and the national issue. Thus, national concern about communist subversion is an issue of the political right and cannot be used to explain the activities of antinuclear leaders on the political left.

As the antinuclear movement changed from one wave to another, the specific issue content of the controversy also changed to become more compatible with the current large issue of national concern. Yearly opinion polls, taken since 1966, have asked people who oppose nuclear power plants for their reasons. Issues of water and thermal pollution and low-level radiation became prominent in the late 1960s with the rise of the Environmental Movement and then passed away with it in the early 1970s (Mazur, 1975). More recent concerns about moving toward a low energy-consumption society, and developing solar and renewable energy sources, reflect public attention to the national energy problem.

Two large issues appear to account for the three major peaks of opposition to fluoridation. Americans feared socialism and communism in the United States, perceiving in them a threat of pervasive central government which would undermine individual liberties. These concerns were intimately connected with the Cold War, and they reached their height in the early 1950s. They reappeared in the mid-1960s and were associated with the presidential campaign of Senator Goldwater. The other national issue was the environment, including concerns over trace "poisons" such as DDT, mercury, and fluorides. Periods of rising opposition to fluoridation—in the early 1950s, 1964-1965, and the late 1960s—coincide with periods of national concern, in Gallup polls, over these two large issues. Environmentalism was not a partisan issue and therefore was politically compatible with the conservative antifluoridation movement as well as the liberal anti-nuclear movement. Concern over the dangers of socialism could be compatible only with a conservative movement.

To explain "small" movements, such as those against nuclear power and fluoridation, as the result of "larger" national issues, such as the environment or energy, may seem a chicken and egg problem, for doesn't each contribute to the other? My argument only has force when the large issue is clearly of much greater concern to the public and the media, and much more overarching, than the particular protest being explained. This is the case in all of the instances I have mentioned here, the larger issues always being of sufficient magnitude to appear on

Gallup polls which ask for the most important issue facing the country, and the particular protests never appearing.

Prediction

It must be clear to the reader that the data base used here is grossly inadequate. The *Readers' Guide* index of mass media coverage totally ignores television and the newspapers, makes no distinction between articles slanted for or against the technology, and allows only the crudest indication of the timing of media exposure. Public opinion polls are erratic in timing, question wording, and even sample design. The measurement of protest activity is the most difficult of all without special resources being alloted to that task. Nonetheless, I believe that the data are sufficient to indicate the benefits of this approach.

If the model of fluctuating opposition described here is dependable, and one can identify the major national issues which drive a controversy, then predictions of future controversy activity ought to be feasible. When this theory was first formulated, a prediction was offered that the energy crisis, a newly emerged national issue, would reinforce the opposition to nuclear power plants (Mazur, 1975). This prediction was considered counterintuitive by many commentators at that time who thought that a pressing need for energy ought to alleviate opposition to an energy-producing technology, but it is now clear that the prediction was correct.

There is a great deal of utility in the prediction/testing approach to social science, so I will make another prediction now, to facilitate the testing of these ideas, recognizing that this attempt may date this chapter very quickly.

It is 1982, three years after the Three Mile Island accident. Barring similar accidents in the near future, the persistence of antinuclear protest in the United States depends on the continued strength of national concern with the energy issue, which can be easily followed by the Gallup poll question about "most important issue." Public concern is quite separate from the underlying reality of the energy issue and the public is losing interest. There are no gasoline lines or heating oil shortages. A few years of freedom from energy worries would remove the driving force from the nuclear power controversy, whereas the persistence or increase in these concerns would increase protest. There is also the possibility that the energy concern will fall away, only to be replaced by another large national issue which would give new force to the controversy. It is impossible to foresee whether or not that will happen, but if it does, then the new issue will appear at that time on the Gallup poll, it must have obvious relevance to nuclear power, and it will have to be an issue of the political left in order to be compatible with the left orientation of the antinuclear movement.

The best candidate for a replacement is the rising public concern over nuclear arms, which may soon register on the Gallup poll of most important problems. However, nuclear weapons and nuclear power reactors have remained separate as issues since the early 1960s, perhaps because some of the major American spokesmen for arms control are supporters of nuclear power (Holden, 1982). My guess is that rising concern over nuclear war will not help the nuclear power protest much but will shift the energies of liberal activists toward the MX and B-1, thus reducing the resources applied against power plants.

References

Ames, Mary. 1978. *Outcome Uncertain.* Washington, D.C., Communications Press.

Carson, Rachel. 1962. *Silent Spring.* Boston: Houghton Mifflin.

Casper, Barry and Paul Wellstone. 1981. *Powerline.* Amherst, MA: University of Massachusetts.

Coleman, James. 1957. *Community Conflict.* New York: Free Press.

Crain, Robert, Elihu Katz, and Donald Rosenthal. 1969. *The Politics of Community Conflict.* Indianapolis: Bobbs-Merrill.

Gerlach, Luther. 1978. "The great energy standoff." *National History* 87(January): 22-32.

_____ and Virginia Hine. 1970. *People, Power, Change: Movements of Social Transformation.* Indianapolis: Bobbs-Merrill.

Holden, Constance. 1982. "Antinuclear movement gains momentum." *Science* 215 (12 February): 878-880.

Kriesberg, Louis. 1973. *The Sociology of Social Conflicts.* Englewood Cliffs, N.J.: Prentice-Hall.

_____. 1981. *Research in Social Movements, Conflicts and Change.* Volume 3. Greenwich, CN: JAI Press.

Lambright, William. 1967. *Shooting Down the Nuclear Airplane.* Indianapolis: Bobbs-Merrill.

McCarthy, John, and Mayer Zald. 1977. "Resource mobilization and social movements." *American Journal of Sociology* 82(May):1212-1241.

McFarland, Andrew. 1976. *Public Interest Lobbies.* Washington, D.C.: American Enterprise Institute.

MacKuen, Michael, and Steven Coombs. 1981. *More than News.* Beverly Hills: Sage.

McNeil, Donald. 1957. *The Fight for Fluoridation.* New York: Oxford University Press.

Mazur, Allan. 1975. "Opposition to technical innovations. *Minerva* 13(Spring):58-81.

_____. 1981. *The Dynamics of Technical Controversy.* Washington, D.C.: Communications Press.

_____ and Beverlie Conant. 1978. "Controversy over a local nuclear waste repository." *Social Studies of Science* 8: 235-243.

Milch, Jerome. 1979. "The Toronto Airport Controversy." Pp. 49-68 in Dorothy Nelkin (ed.), *Controversy: Politics of Technical Decisions.* Beverly Hills: Sage.

Mitchell, Robert. 1980. "Public opinion and nuclear power before and after Three Mile Island." *Resources* 64(Jan.-April):5-7.

Mueller, John. 1968. "Fluoridation attitude change." *American Journal of Public Health* 58:1876-1880.

Nader, Ralph. 1965. *Unsafe at Any Speed*. New York: Grossman.
Nelkin, Dorothy. 1974. *Jetport*. New Brunswick, NJ: Transaction Books.
_____. 1979. *Controversy: Politics of Technical Decisions*. Beverly Hills: Sage.
Nichols, David. 1974. "The associational interest groups of American science." Pp. 123–170 in Albert Teich (ed.), *Scientists and Public Affairs*. Cambridge, MA: MIT Press.
Obershall, Anthony. 1973. *Social Conflict and Social Movements*. Englewood Cliffs, NJ: Prentice-Hall.
Tukey, John. 1977. *Exploratory Data Analysis*. Reading, MA: Addison Wesley.

Chapter 3

Biotechnology and Unnatural Selection: The Social Control of Genes

Sheldon Krimsky

In his book, *The Second Genesis,* Rosenfeld (1975:32) introduced the term biosocioprolepsis or BSP. Building on the term prolepsis meaning anticipation, the word signifies the anticipation of biology's impact on society. The goal for BSP, according to Rosenfeld, is achieved "by projecting our imaginations ahead into our possible choice of social futures, we try to anticipate the dangers inherent in biomedical advance, and to forestall them by our foresight."

The social control of technological change has not progressed in this fashion. With very few exceptions, the pathological effects of technology have been controlled subsequent to their appearance in the industrial and domestic sectors when significant damage has already been done. Rarely are we faced with an opportunity of establishing safeguards for a technological revolution during its embryonic stages before it has become calcified into our economic system. That opportunity presents itself to us as molecular genetics is brought from the scientific laboratory to industry. But having a new term like BSP, a heightened consciousness about technology's dual edge, or critics who can conjure up prophesies of microbial chaos and humans fouling up evolution is not sufficient to

This chapter is an expanded version of a talk delivered to the American Association for the Advancement of Science on 7 January 1982, subsequently revised for publication in *Environment,* August 1982.

provide the social guidance needed to intercede in what has been termed technology's autonomous path (Winner, 1977). We need institutional mechanisms to monitor, measure and test hypotheses related to the impacts of biotechnology on culture and ecology. The progress toward which we strive in exploiting nature's genetic secrets must have a counterpart in the progress we exhibit for developing social instruments of assessment.

For ten years, concerns raised about the revolution in molecular genetics have focused mainly on the problems of potential biohazards arising from laboratory experiments. The response to a cornucopia of conjectured risks of gene splicing has not been trivial. New relationships have been created between science and the broader society which supports its activities. These relationships are not only unique to the field of biology, but they are unprecedented for the entire enterprise of science. The changes that have taken place include: establishment of a federal office to issue guidelines and oversee the risk assessment for gene-splicing experiments; creation of local Institutional Biosafety Committees with community representation; enactment of laws by nine local governments and two states that regulate the use of recombinant DNA (rDNA) technology for research and commercial institutions (Krimsky, 1982; Dutton and Hochheimer, 1982).

The most noteworthy reform established during the decade made biologists who were engaged in rDNA research accountable to other individuals or institutions for the safety practices in their laboratories. Despite fears by some biologists, the involvement of non-scientists in the rDNA episode has not impeded scientific inquiry in any significant way (Singer, 1979; Setlow, 1979). New institutional mechanisms were developed to respond to a crisis in science over the safety of research. However we may interpret their effectiveness, the institutional forms reflect a greater responsibility of science and government to society. A system of social guidance, steered by scientists, but open to public involvement, undertook the difficult task of trying to assess the laboratory hazards of new technology. It was a rare opportunity for scientists to test their powers of BSP. But the preponderance of attention given to laboratory safety has masked other vitally important societal concerns pertaining to the commercial and military applications of genetic technology. My purpose in this chapter is to draw attention to the potential impacts of biotechnology that take us beyond the inadvertent creation of hazardous chimeric organisms.

The issues I shall raise about genetic technology are more fundamental than a list of actual or hypothetical concerns about its social, economic or environmental impacts. I inquire whether social institutions are in place that can address actual or potential problems associated with new developments in genetic technology, and whether our current institutions are appropriate to meet the demands of the problem. To the reader unfamiliar with biotechnology and rDNA molecule technology in particular, the following sections are offered as a primer to the field.

Rudiments of Biotechnology

The term biotechnology in its broadest sense means the application of biological processes for human purpose. The use of microorganisms to make beer and bread has been traced to antiquity (Demain and Solomon, 1981). In the 1940s the pharmaceutical industry began using microorganisms to produce antibiotics. However, in the last decade a substantial leap has been made in the commercial applications of simple life forms. This revolution is characterized by the expression genetic technology. All life forms from the simplest organisms such as bacteria and yeasts to higher mammals are made up of cells as the basic biological unit. Each cell consists of a set of instructions contained in discrete packets called genes. The genes are composed of threadlike molecules called DNA (deoxyribonucleic acid) grouped together in units called chromosomes. Bacteria consist of 2-3,000 genes on a single chromosome. Human cells contain 46 chromosomes with over 100,000 genes. The genetic instructions in the cell determine its growth and structure including its primary products—proteins.

Genetic technology refers to those processes through which the genetic instructions of a cell in the animal, in the test tube or in culture, can be controlled, manipulated, or transferred to other cells. For several decades scientists have been able to produce genetic changes in cells by the use of radiation, infection by viruses, chemicals, and exposure to pure DNA. Once the genetic changes were made, cells could be selected out for a particular purpose, such as hardier strains of wheat, or fungi that produce greater yields of antibiotics. In 1973 scientists discovered a method of wide applicability for transporting individual genes from one cell to another of virtually any species. This technique of recombining genes (recombinant DNA) meant it was possible to reprogram microorganisms to synthesize proteins that were completely foreign to them.

A typical recombinant DNA experiment involves three basic steps: (1) extracting a gene segment from a donor cell; (2) joining the gene in a test tube with a carrier DNA molecule (the foreign gene attached to the carrier DNA molecule is called the recombinant DNA molecule); (3) transporting the recombinant DNA molecule into the host cell.

The power of this technique compared to previous forms of genetic engineering is that it established genes as completely fungible elements capable of being transported between organisms however distantly related. Furthermore, it is a great advance over hit and miss methods of genetic engineering by mutation and selection.

When foreign genes are carried into bacteria the progeny cells of the microorganisms will get copies of the new gene. Under such conditions the foreign gene is said to be cloned or duplicated by the genetic apparatus of the cell. In addition to being able to produce large amounts of pure DNA through gene splicing, it also can be used to synthesize the protein encoded by the foreign DNA. Thus, by introducing the appropriate genes into bacteria these organisms can be transformed into biochemical factories for synthesizing substances useful to medicine, industry and agriculture.

Medical and Industrial Applications of Genetics

In its widely circulated study, *Impacts of Applied Genetics,* the Congressional Office of Technology Assessment (OTA, 1981) cited five areas where rDNA will have the greatest impact: pharmaceuticals, chemicals, food, agriculture and energy. The respected financial weekly Barrons (Nossiter, 1982:8) reported that more than 100 specialized companies are trying to capitalize on the applications of gene splicing to these fields responding to markets estimated to pass $3 billion by 1990 (Patterson, 1981:66).

The earliest and most widely publicized application of recombinant DNA techniques is in the production of medically important proteins for use in research and the treatment of disease. Among the products currently being manufactured or still in the development stage are human insulin, animal and human growth hormone, clotting factors, antibodies, vaccines and interferon. From a scientific standpoint, virtually any human protein is subject to bacterial biosynthesis if the genes which encode it can be implanted and made to function in the bacterial environment. In vaccine production, the use of gene transplantation has made it possible to manufacture large quantities of non-virulent, non-selfreplicable segments of a virus that can be used to immunize a host. Two European companies are purported to be the first in the world to market a product manufactured by genetically engineered bacteria. Burroughs Wellcome of London and Intervet International (a subsidiary of the Dutch chemical firm Akzo A.V.) are manufacturing a vaccine to protect piglets and calves from scours disease (infectious diarrhea).

In agriculture, large investments have been made in rDNA molecule technology with the hope of producing a new class of "superseeds" even hardier and more fertile than those associated with the first "green revolution." Molecular genetics promises to provide the knowledge base underlying the genetic determinants of high yield strains. Among the most hotly pursued aspirations of rDNA technology applied to agriculture is in the area of nitrogen fixation. Certain bacteria and blue-green algae can transform free nitrogen from the atmosphere into ammonia which plants need for their growth. The bacteria which perform this function live at the root nodules of legumes such as soybeans and peanuts. But there are many valuable crops such as wheat, corn, and cereal grains, for which bacterial nitrogen fixation does not occur. For these plants, yields have been improved through the use of chemical fertilizers which have introduced many environmental problems.

The nitrogen fixation process is associated with a discrete set of genes (nif genes) in the bacteria. The new-found ability to move genes between species has prompted three basic research strategies for improving on nature's use of nitrogen fixation: (1) increase the efficiency of nitrogen fixation for the plant-bacterial systems that currently possess this capacity; (2) genetically transplant the nif genes into new microorganisms; (3) genetically transplant the nif genes directly to plant cells making them self-fertilizing.

The food industry also has a serious eye on genetic engineering. OTA (1981: 107) cites two ways that microbial activity is used in food processing: (1) inedible biomass is transformed by microorganisms into food for human consumption or animal feed; (2) organisms are used in food processing either by acting directly on food or by providing material that can be added (such as enzymes and vitamins).

The prospect of revolutionizing the sweetener technology by developing cheaper methods for manufacturing pure fructose has encouraged an $8 million investment in the Cetus Corporation by the Standard Oil Company of California. The U.S. fructose market is estimated to be $11 billion a year. One of the first food-processing products manufactured by rDNA technology is rennin, an enzyme that turns milk into cheese. Collaborative Research with investments from Dow Chemical Company genetically engineered a yeast to express the enzyme.

In the energy field, bacterial strains are being sought which can economically convert agricultural and forest biomass into liquid fuels. Patents have been granted for genetically engineered microorganisms that can detoxify hazardous wastes or degrade oil spills. Currently, bacteria play a minor role relative to chemicals in the multibillion dollar insecticide industry. Approximately a dozen biological agents have been registered in the U.S. as pesticides. Scientists are looking for ways to improve the potency of bacterial strains on their pest targets by increasing genetic determinants of the toxins that destroy insect pests. Meanwhile, OTA (1981:89) projects there will be a revitalization of biotechnology in the chemical industry. It is expected that bacterial fermentation of certain substances will be substituted for selected chemical conversion chains that are part of the overall manufacturing process.

Considered along with computers as having enormous growth potential in the next several decades, biotechnology has touched off a major investment revolution. Most leading chemical companies are currently involved in genetic engineering either in-house or through patent and marketing agreements with smaller firms and universities (Fox, 1981:17).

The following are some highlights of the investment activity that had taken place by 1981 in the field of biotechnology. The Schering-Plough Corp. owned 16% interest in the Swiss bioengineering firm, Biogen. American Cyanamid had 20% equity in Molecular Genetics, Inc. The National Distillers and Chemical Corporation owned 11% of Cetus Corporation. Koppers Company, Inc., owned 48% of Genex Corporation. Dow Chemical invested $5 million for 5% of the common stock of Collaborative Research. Standard Oil had a 17% investment in Cetus Corporation. E.I. duPont de Nemours & Company was operating its own genetics research facility and paid Harvard University $6 million for the exclusive rights to produce and market products that were derived from the university's genetics discoveries over a period of five years. Eli Lilly was involved in joint ventures with Genentech and entered into a long term agreement with International Plant Research Institute of California on improving plant yields. Phillips Petroleum paid $10 million constituting 35% equity in the Salk Institute Biotechnology Corporation which used to be a wholly owned subsidiary of the Salk Institute in La Jolla, CA. Mallinckrodt, Inc., a chemical company in St. Louis, invested $3.88 million in Washington University for research in hybridoma technology.

The Massachusetts General Hospital (MGH) signed a contract with one of the world's largest drug and chemical firms, Hoechst A.G. of West Germany. Under the plan Hoechst will provide MGH with $50 million over a decade to launch a major research program in genetic engineering in return for patenting and marketing rights. The Whitehead Foundation provided a $5 million operating budget and an initial $20 million grant for the construction of an independently run Whitehead Institute for research in molecular biology with cooperative ties to M.I.T. In the event of Mr. Whitehead's death, the new institute will obtain an additional $100 million under the agreement.

This inventory is only meant to be indicative of the investment activity in 1981; it is neither comprehensive nor suggestive of future trends. But it does help us understand the economic forces that are driving the application of rDNA technology into the commercial sphere.

Against the current of the extraordinary investment fever, there are some who question how this technological wonderland of genetic chemistry will affect our society. A former chairperson of both the Recombinant DNA Advisory Committee (RAC) and the House Subcommittee on Science, Research and Technology, Ray Thornton (1981), made the following poignant remark to the RAC: "Human experience has shown that any tool powerful enough to produce good results of sufficient importance to shake Wall Street and offer hope of treating diabetes is also powerful enough, wrongly used, to produce bad results of equal consequence."

In the context of these remarks I shall address five areas of social concern for the field of biotechnology.

Harnessing rDNA Technology

Now that molecular biology has important social applications why doesn't our government establish priorities for harnessing gene-splicing technology in the interest of the greatest number of people? Is there any justice in allowing the free market to determine whether and to what extent gene splicing improves people's living conditions by determining what products are introduced into the market place? Three arguments have been advanced in support of a strong governmental role in exploiting the social uses of genetic engineering.

Argument 1. Since public monies were the principal source of funding through which rDNA methods were developed, the public sector should play a major role in directing its use. A corollary to this position states that the public is entitled to a return on its investment and should control the patents on products and processes that grow out of federally funded research.

Argument 2. If social priorities are not set for the use of rDNA technology then the public will miss out on important applications which private markets will not find profitable to pursue. A case in point is "orphan drugs" which illustrates the need for governmental involvement in the development of pharmaceuticals. The drugs are so named because they cannot find a parent company who will invest in their manufacture. The markets for the drugs are too limited to provide a satisfactory return on investments. Few question, however, the responsibility of society to make available for clinical use non-profitable drugs if they are effective in aiding even a small number of patients.

Argument 3. When the fruits of rDNA technology are realized, such as in agriculture, it is the responsibility of government to guide the benefits so that they are at least shared equitably and at most shared in a manner that narrows distributional gaps. In the case of agricultural impacts, a guidance system can insure that small farmers are not disadvantaged from new strains of genetically engineered seed stocks, that the consumer gets a better quality product at a more reasonable price, and that environmental health is not traded off for higher rates of return.

For the purpose of this discussion I shall suspend judgment on the cogency of the arguments. They form an essential part of the background criticism that has been raised against the fledgling gene-splicing industry. Hundreds of biotechnology firms have surfaced in a highly competitive marketplace with their own sets of agendas and perceptions of social needs. Returning to my initial query: Are there institutions which can establish priorities for harnessing rDNA technology? What assurances are there that private and public investments in the field of biotechnology get channeled into uses that are responsive to distributional inequities.

Currently, no single institution has the authority to set and implement priorities for the application of gene-splicing methods to industrial, agricultural, or clinical areas. Moreover, there is little, if any, precedent in this country to guide the development of a technology of such broad scope. While we have governmental institutions for setting research agendas, assessing and controlling technological impacts, and overseeing targeted programs in applied technology, private markets are fundamentally responsible for what gets produced, in what order and toward what end.

In theory at least, the Office of Technology Assessment possesses an excellent vantage point from which to establish a set of priorities for developing rDNA technology. But on the basis of its evolving role over the past several years, which excludes advocacy of particular policies and actions, it is highly unlikely that OTA would be the body to establish a hierarchy of needs from which to develop a strategy for extracting social benefits from the technology. Different agencies of government such as the Department of Energy (DOE), the Food and Drug Administration (FDA) and the Department of Agriculture (USDA) will set their own agendas. However, the public has little access to how these agencies establish their individual priorities.

Recently, the Plum Island Animal Disease Center (USDA) and the bioengineering firm Genentech, Inc., entered into a cooperative agreement to produce a vaccine for foot and mouth disease. For many countries outside the United States, the highly infectious foot and mouth virus is responsible for substantial losses in beef stock and milk production. Fortunately, North American agribusiness has been spared the disease for many decades because of strict beef import controls. It is argued that American consumers could benefit from a worldwide eradication of the disease. For Genentech, the carrot in this public-private partnership is the right to foreign markets for vaccine sales, a sizable benefit for a small firm that entered the vaccine research program in its final stages.

Broader public input for setting agency priorities could come from Congress through its appropriate subcommittees on science and technology. In its report *Impacts of Applied Genetics,* OTA (1981:10–12) generated several options for Congress to consider with regard to promoting advances in biotechnology. The options included: establishing a funding agency in biotechnology; creating federally financed research centers in universities; providing tax incentives to expand the capital supply to small high risk firms; improving conditions under which U.S. companies collaborate with academic scientists; mandating support for specific research programs. Each of these recommendations has been used in the past to stimulate or set a direction for the development of innovative technologies. But the approach taken by OTA does not address the question of setting overall priorities for the utilization of rDNA technology on the basis of social needs. Precedents in this country run against this type of endeavor which might be termed national technology planning.

In contrast, the federal government has taken an active role in shaping the direction and quality of research, both targeted and basic, through funding mechanisms (Cooper and Fullarton, 1978). It is estimated that about two thirds of all the basic research carried on in the U.S. is federally supported. But the public sector role has been minimal to almost nonexistent in directing the application of technologies. It is widely assumed that social needs will be more effectively revealed through market forces. With due respect to the sudden growth of economic fundamentalism, there are many areas where the assumption fails miserably. A former member of the White House Office of Science and Technology Policy and an astute observer of genetic technology offered this prognosis (Omenn, 1981:44).

There is certain to be a shake-out in biotechnology over the next five years or so, and the determinants of success are likely to be related more to business strategies, shrewd management, and high quality control, assuming a strong base of laboratory talent. Chemical and agricultural products will be marketed in short order, if they are economically competitive, but hormones, drugs, and vaccines must undergo complex and expensive clinical trials. There can be little doubt that the Wall Street criteria will be applied: earnings, growth, profitability.

Since gene splicing is expected to introduce innovations in several commercial fields, if a hierarchy of needs is conceded to be desirable, it seems more reasonable that it be achieved within specific areas, such as vaccine production, chemical processing, biomass conversion and agricultural products. Federal agencies such as FDA, DOE and USDA could set priorities in biotechnology with appropriate inputs from the public. It has also been suggested that "we might set as a national goal the conversion of an economy based on fossil fuel to an economy based on microbial fermentation" (Lewis, 1981:46). A national effort, on the scale of the space program, that promotes the development of inexpensive, renewable energy sources from biotechnology would improve significantly the public's confidence in science and technology.

In the pharmaceutical field, who decides what gets manufactured first, bovine growth hormone to fatten up cattle or human growth hormone as a replacement therapy for a genetic defect? Should a vaccine for malaria get precedence over one for Herpes virus? Other factors besides profit and social demand will invariably enter into such determinations, such as how far advanced the state of knowledge is toward a solution of a particular genetic engineering problem. Considerations also include what sources of private and public capital are available for specific product development. Products with low market potential are likely to be left behind. There are no established institutions or advocacy networks through which the public sector can make its voice heard on priorities in technological innovation. Yet the public has been promised so much from biotechnology in such a short time that federal responsibility and initiative in supporting a development program deserves careful consideration. The first step is to recognize the legitimacy of the citizenry in steering a technology. The next step is to develop avenues of participation and social guidance mechanisms.

Technology Transfer to the Third World

Among the important applications of biotechnology, some will eventually be exported to third world countries. What responsibility do we bear in the transfer of this technology to the developing nations? On one hand, the industrialized world must find ways to share the positive fruits of genetic engineering with developing nations without destroying their unique cultural forms, neglecting their capacities for self-determination, or disregarding the needs of their economic systems. On the other hand, we bear an obligation to prevent the export and development of products and processes that we determine to be unfit for ourselves or which would be unsuitable for the cultural patterns and technological development of the country in questions.

Very soon after plasmid-mediated transfer of chimeric genes was discovered, its commercial applications were being seriously investigated. One scientist, who was employed by a major U.S. corporation, saw in genetic engineering a solution to the problem of world hunger. This scientist was planning to construct a plasmid with genes from *Pseudomonas* that code for enzymes with cellulose-degrading properties. He considered cloning the plasmid with the cellulose-degrading genes in *E. coli*. His logic for using this organism to alleviate world hunger was as follows: suppose that vast populations in underdeveloped food-scarce countries could have their intestinal flora transformed or replaced with the new cellulose-degrading *E. coli*. With their new intestinal flora, these individuals could presumably obtain caloric value from vegetation that is plentiful and inexpensive, but under the present circumstances nutritionally useless to them. After being advised by a scientific colleague that cellulose-degrading *E. coli* in the gut could eliminate any roughage in the digestive tract and thus increase the rates of certain disease correlated with low fiber diets (obesity and bowel cancer were cited) the investigator gave up his plan (Krimsky, 1982:117-119).

I offer this story not to emphasize the potential hazards of such a scheme nor to question the motives of responsible scientists, but to remind us that the idea was being considered for use exclusively in the poorer nations of the world. Currently, transnational corporations market and export products to developing countries which are either prohibited or severely restricted for domestic use. Recent publications and television documentaries have illustrated these problems for the export of pesticides and pharmaceuticals (Schulberg, 1979; Weir and Shapiro, 1981; PBS, 1981). There are no U.S. federal agencies with authority to prohibit the export of domestically banned products even when they are manufactured in this country. Under the Federal Insecticide, Fungicide and Rodenticide Act (FIFRA) and Toxic Substances Control Act (TSCA) there is a statutory obligation to inform foreign governments of regulatory actions taken against specific products. Also, under TSCA, companies are required to make toxicity information on chemical products available to foreign import countries. Notwithstanding the legality of the transactions, and those cases of domestically-banned exports whose benefits to the import country clearly outweigh their adverse effects, there remain many areas where the ethical practices of the exporting firms are of a questionable nature.

Just prior to leaving office President Carter signed an executive order (15 January 1981) that placed export controls on extremely hazardous substances deemed a substantial threat to human health, safety, or the environment. Once classified, these products would require an export license. Carter's order took two years to develop. It was in effect for only a month when it was revoked by President Reagan (Shaikh and Reich, 1981).

Within a couple of years the first commercial products of rDNA technology will be reaching consumers. It is reasonable to expect the problem of questionable exports to be compounded as biotechnology's pioneers seek world markets. For example, since pesticides contaminated with dioxins are already being sold to some developing countries, there is also a market for the biological antidote — organisms genetically engineered to degrade this class of pesticides. The media have recently reported the development of a new genetically engineered bacterium which degrades the herbicide 2,4,5-T. Its creator, who also developed an oil-eating bacterium, was quoted as saying: "If you use 2,4,5-T to kill weeds one year and then apply these microorganisms to clean up the 2,4,5-T . . . their number will die out drastically when there is no more of the chemical to eat" (Chakrabarty, 1981). Under such ideal but unrealistic conditions, the microbes will not mutate, establish themselves in new niches, nor adversely affect the microflora of the land areas on which they are sprayed. Less dubious assumptions have been responsible for creating havoc in sensitive ecological systems.

Our current experience with exports of hazardous materials and the transfer of certain inappropriate technologies to industrially underdeveloped agrarian societies leads me to the conclusion that unless more responsible institutions are created, similar mistakes will be made when the fruits of rDNA technology become realized. The widely used justification that importing countries are free and willing to buy products of dubious value fails the test of moral reciprocity. The Carter executive order was a positive step toward a global responsibility for our exported products. With that order rescinded, an additional burden is placed on scientists and the public health community who are familiar with the deleterious effects of chemical or biological exports to inform the recipient countries and the World Health Organization before rather than after severe injury or environmental damage is incurred.

Secondary Impacts

The products, processes, and industries that eventually emerge as a consequence of the commercial applications of molecular biology will undoubtedly exhibit unintended secondary impacts. For example, if a vast array of new or old pharmaceutical products are manufactured at low cost (including antibiotics and insulin), what effect, if any, will that have on drug overuse? Harvard biologist Ruth Hubbard (1977:165-169) assessed the use of rDNA technology for the production of insulin at a meeting of the National Academy of Sciences: "given the history of drug therapy in relation to other disease, we know that if we produce more insulin, more insulin will be used, whether diabetics need it or not."

If agricultural plants are engineered to be resistant to herbicides, will that stimulate a greater use and dependency of chemicals in agriculture? If microbes are developed which can degrade herbicides containing dioxins, will that justify the removal of restrictions on this class of potent chemicals when they are used in conjunction with their biological antidote?

It is easy to raise hypothetical cases. I do not pretend to have any skills as a technological forecaster. I use these not-so-implausible cases to reintroduce the theme of my inquiry: Are there social guidance systems through which society can anticipate, or at least keep track of the secondary consequences of major technological innovations?

In the late 1960s, the term technology assessment became a part of our policy vocabulary. The National Environmental Policy Act established a requirement for environmental impact assessments for many government supported projects. Even the NIH guidelines for recombinant DNA research were issued with an environmental impact statement (USDHEW-NIH, 1977). Some laws and institutions are already in place to respond to the potential impacts of biotechnology. None, however, are equipped to provide continuous monitoring and assessments of the full range of expected commercial uses of gene splicing, and other tools of biotechnology such as cell fusion, over a period of years.

In principle, the Office of Technology Assessment is well suited to carry out this function because it has been able to assemble highly trained interdisciplinary teams of scientific and policy experts. But as an agency of Congress, OTA will undertake studies when there is sufficient congressional support. In its recent report on applied genetics, OTA chose to place considerable weight on the positive social outcome of genetic engineering including increased manpower needs, new products, and improved yields in agriculture. But only a feeble effort was made in this study to evaluate potential adverse social or ecological consequences of industrial microbiology. No research program or framework for long-term assessment is offered.

A second study by OTA started in the fall 1981 evaluates the competitive position of the United States in the global biotechnology field. In the scope of project activities there is a striking absence of any reference to the assessment of secondary impacts. For OTA, technology assessment in the field of biotechnology has come to mean promotion, pure and simple.

Individual agencies also bear responsibilities under legislative mandate to evaluate products of rDNA technology. But evaluations of this nature are spotty and restricted in scope. The FDA is primarily concerned about the efficacy, purity, and the side effects of a new drug, but cannot rule on its broader social manifestations. The Environmental Protection Agency (EPA) has no authority to restrict the use of pesticides on the grounds that they might exacerbate the decline of small farms. (See the chapters by Dorner and Thiesenhusen for the effects of technology on small farms.) Under a variety of statutes EPA does have authority over the release of hazardous materials into the environment. However, no regulations currently exist or are anticipated that restrict the release of biological agents into the environment with the sole exception of biological pesticides which must be evaluated for safety and efficacy according to the Federal Insecticide, Fungicide, and Rodenticide Act (FIFRA) before they are registered. While initially planning to issue regulations for infectious wastes under the Resource Conservation Recovery Act, EPA has changed its approach under the Reagan administration. Its current effort is directed at publishing an infectious waste management plan to help industry on a purely voluntary basis become acquainted with accepted practices for disposing of biological wastes.

It is clear that biotechnology is today where the petrochemical and nuclear technologies were forty years ago. Our experience in these fields should not be neglected. Without adequate guidance systems the social consequences of technology come without advance warning and in a form in which effects are all too often irreversible.

rDNA and Biological Weapons

The feasibility of creating biological weapons with rDNA technology was on the minds of those who attended the Asilomar conference in 1975 where scientists from 15 nations met to discuss the science and potential risks of gene splicing. Despite expressed concerns by some participants, the issue was kept off the agenda by the conference organizers for fear it would interfere with the primary goal of reaching a consensus on laboratory biohazards (Krimsky, 1982:106). Nevertheless, one of the three working panels at Asilomar concluded its report on the assessment of risks with the following admonition (Plasmid Working Group, 1975:19):

> We believe that perhaps the greatest potential for biohazards involving genetic alteration of microorganisms relates to possible military applications. We believe strongly that construction of genetically altered microorganisms for any military purposes should be expressly prohibited by International treaty. . .

Just a few months prior to Asilomar, the United States Senate ratified the articles of the 1972 Convention on the Prohibition of the Development, Production, and Stockpiling of Bacteriological and Toxin Weapons. The Convention's articles were put into force in this country by March 1975. Nearly a hundred countries had already pledged "never in any circumstance to develop, produce, stockpile or otherwise acquire or retain (1) microbial or other biological agents . . . whatever their origin or method of production, of types and in quantities that have no justification for prophylactic, protective or other peaceful purposes; (2) weapons, equipment or other means of delivery designed to use such agents or toxins for hostile purposes or in armed conflict" (Bacteriological and Toxin Weapons, 1975:5).

According to Article XII of the Convention, a review is to take place five years after it has been in force to examine the relevance of new scientific and technological developments. A United Nations review committee report issued on March 1980 concluded that biological materials constructed by rDNA techniques were unambiguously covered in the Convention's language (Preparatory Committee, 1980:7). The committee also assessed the potential use of rDNA for creating biological weapons. The prospect of developing fundamentally new agents or toxins with rDNA technology was viewed as a problem of "insurmountable complexity." The committee saw little incentive for such efforts since "naturally-occurring, disease-producing micro-organisms and toxins already span an exceedingly broad range, from some which are extraordinarily deadly to others usually producing only temporary illness." However, the committee considered more probable the use of rDNA techniques to improve the effectiveness of existing biological warfare (BW) agents.

What assurances does the public have that present or future administrations will adhere to the articles of the Convention? What institutions are currently available to provide the public with information on biological research programs carried out within or funded by the Department of Defense? Is there a clear distinction between offensive and defensive biological weapons? Does it even make sense to speak about defensive biological weapons? What would they be and is their production restricted by the Convention?

As an example, might the army want to clone toxigenic genes into wild type *E. coli* to assess the potential use of gene splicing by a hostile state or terrorist group? Should our defense establishment be funding projects to make vaccines for pathogens that do not currently exist? According to a 1980 report, DOD has expressed an interest in assessing whether genetic engineering can be used to make new biological weapons (DOD, 1980, sec. 2:4).

> New threats may be opened up by various technological and scientific advances. As examples, recombinant DNA technology could make it possible for a potential enemy to implant virulence factors or toxin-producing genetic information into common, easily transmitted bacteria such as *E. coli*. Within this context, the objective of this work is to provide an essential base of scientific information to counteract these possibilities and to provide a better understanding of the disease mechanisms of bacterial and rickettsial organisms that pose a potential BW threat, with or without genetic manipulation.

To improve its understanding of these possibilities the U.S. Army has begun some rDNA work. Its medical Research Institute of Infectious Disease received permission from the RAC to clone *Pseudomonas* exotoxin in *E. coli (Science,* 1980a). The Army Medical Research and Development Command advertised in *Science* for proposals on the introduction by rDNA methods of the human nervous-system-gene acetylcholinesterase into a bacterium *(Science,* 1980b).

The expressed purpose of the research is to develop an effective antidote for nerve agents manufactured by the Soviets which are extremely potent cholinesterase inhibitors. The Army's interest in cloning the enzyme is to obtain a sizable quantity of high purity material so that its physical and biochemical properties can be studied.

What windows of accountability exist between the military and the public on the uses of genetic engineering? How can public skepticism be turned into public confidence? Presently, there are three institutional responses that serve to build public confidence: the 1972 Convention previously mentioned; a federal law requiring the DOD to describe its obligated funds in chemical and biological research; and the NIH guidelines. The second institutional response gives Congress and the public more direct access to the chemical and biological research carried on by the military. According to Public Law 91-121 passed in 1969 and amended in 1975 (P.L. 93-608) the DOD is required to submit an annual report to Congress that explains each expenditure in its chemical warfare and biological research programs including those designed for "development, test and evaluation and procurement of all lethal and nonlethal chemical and biological agents."

The third instrument of social accountability is the NIH guidelines. A memorandum from the Undersecretary of Defense dated April 1, 1981 states that "all DNA activities funded by DOD, whether in-house or by contract or grant, will be conducted in full compliance" with the NIH guidelines (Wade, 1981). The ruling specifies the use of institutional biosafety committees and requires that a complete file of each research project be maintained for public scrutiny at the U.S. Army Medical Research Institute of Infectious Diseases.

It appears then, that if the Army wishes to clone toxigenic genes into wild type *E. coli,* current DOD policy requires that these experiments must first receive approval from the RAC and subsequently be registered. There is no reason to believe that DOD will not adhere to the principles of the Convention. At issue here are the bridges of confidence that must exist between the public and the military in the context of possible alternative interpretations to the language of the treaty.

For example it is not clear whether the Convention language prohibits the construction of new or improved pathogenic strains of bacteria if the putative interest in such agents is either to determine their strategic capability for military and civilian populations or to aid in the manufacture of a vaccine against them. On the conjecture that some nation has the capability to construct a new strain of a virulent organism that could serve as a BW weapon, another may decide to construct it first in order to develop a vaccine. Our defense establishment becomes particularly vulnerable to this type of thinking when fears are raised about the escalation of Soviet chemical and/or biological weapons activity (Marshall, 1981).

Richard Goldstein, a molecular geneticist at the Harvard Medical School and member of the RAC, argued that the DOD can construct altered forms of virulent pathogens for biological warfare and still be conforming to the principles of the Convention "if the rationale is that the work is being done for prophylactic, protective and peaceful purposes." Under the Convention's rules Goldstein believes that a country can work with superpathogens, produce vaccines against such agents, and develop the dispersal systems for delivery in order to defend itself against a BW system. "[H]aving perfected such systems under the blessings of the Convention (i.e. for defensive purposes), DOD in reality has a cleverly disguised offensive capability for biological warfare" (Goldstein, 1982).

Some time in early 1982, a confidential proposal was sent by the Army to the National Academy of Sciences which described classified experiments it was prepared to fund. According to a report in the British science journal *Nature* (Budiansky, 1982:615) the experiments included "the possible offensive uses of recombinant DNA technology in biological warfare, ostensibly for the purpose of better understanding how to defend against them." The contents of the proposal leaked out and the military use of rDNA molecule technology became a subject at the June 1982 RAC meeting. Richard Goldstein and former RAC member Richard Novick of the Public Health Research Institute of New York City proposed the following amendment to the NIH guidelines:

> The use of recombinant DNA methodology for the purpose of development of microbial or other biological agents or toxins as biological or chemical weapons is prohibited as consistent with the spirit of the 1972 Biological and Toxin Weapons Convention.

The RAC defeated the proposal and instead passed a motion to remind the Director of NIH that the Convention prohibits the use of rDNA methods to produce agents not used for "prophylactic, protective or other peaceful purposes." Meanwhile, it has become clear that there will be no public oversight of the classified research funded by the DOD assessing the potential of rDNA for biological weapons which could involve the construction of pathogens totally unique to the ecosphere.

Human Genetic Engineering

The discovery of gene splicing as a tool of scientific inquiry received considerable media attention because of the initial concerns about producing hazardous organisms. But even as these issues were debated sectors of society began to draw attention to human genetic engineering. States and local communities that took up the regulatory issues were confounded by the ethical ones, if they considered them at all. The exception is Waltham, Massachusetts which passed a law that forbids the use of humans in rDNA experiments. The human-experiment provision, tagged on to its ordinance regulating rDNA activities, did not result from a broad community debate, but was prompted by a single member of the city council.

What safeguards are in place to protect society from the potential misuse of biotechnology in human genetic engineering? Are there any ethical thresholds which should be considered when applying genetic engineering to the treatment of disease or in conjunction with other reproductive technologies? Should limits be set beyond which clinical research in human genetics becomes impermissible?

It is not difficult to conjure up insidious forms of human genetic manipulation as in cloning of an individual. Nor is it problematic to think of humane uses for genetic therapies. But there is a vast middle ground for the human applications of genetics for which a consensus does not exist among scientists, ethicists, members of the religious community and the general public. Some view alteration of germ line cells as morally reprehensible. Others argue that we bear just as much an obligation to eliminate from the gene pool the determinants of Tay Sachs and sickle-cell anemia as we have to eliminate smallpox from the planet. Germ line cell surgery may be the only way to achieve such a goal.

In 1977, at the National Academy of Sciences' Academy Forum devoted to genetic engineering, a scientist tried to put into perspective both the lofty claims and the exaggerated fears being expressed about rDNA technology. Questioning the promise of genetic surgery he said: "How about a thalassemic? Are we going to drain his marrow out, then culture his cells, get DNA in (the cells) and put (them) back in (the person)? Quite frankly I would rather be a thalassemic than have that happen to me" (NAS, 1977:170).

However incredulous the implantation of genetically engineered cells may have appeared at the time, just three years later a UCLA investigator performed a remarkably similar procedure in what has been deemed the first human genetic engineering experiments. These experiments were conducted in Israel and Italy using Italian and Israeli subjects, who voluntarily consented to participation. The subjects in question were suffering from a life-threatening blood disease (beta thalassemia major) in which the bone marrow cells are unable to produce normal hemoglobin because of defective genes.

The UCLA investigator removed bone marrow cells from the patient and exposed the cells to normal genes for making hemoglobin. The human genes were cloned by rDNA techniques. The genetically engineered bone marrow cells were reinjected into the patient in the hope that they would prosper and produce blood cells with normal hemoglobin. The individuals treated by this procedure were experiencing the final stages of the disease and given the prognosis of a limited life expectancy (Schmeck, 1981:20E).

Similar protocols were not approved for experimentation at UCLA by that institution's human experimentation committee. After holding the application for fourteen months, the UCLA Human Subjects Committee was unwilling to approve the experimental procedure on the grounds that there was insufficient evidence of its success in animal systems.

What can be said about the current institutions available to handle such questions? Committees for the protection of human subjects, whose operating procedures are defined in federal guidelines, issue independent judgments at each institution. But human genetic engineering represents a quantum leap in the use of humans as experimental subjects. There are more issues at stake than the protection of the rights of privacy and well being of individual subjects.

In cases where the genetic engineering is performed on the fertilized egg *in vitro* (combining genetic engineering with *in vitro* fertilization) review by human subject committees is not required. It is conceivable that the Recombinant DNA Advisory Committee could address the issue of human genetic manipulation. But its current charter and membership is designed to keep RAC's attention exclusively to biohazards and away from ethical problems.

There is another institutional process for reviewing human genetic engineering experiments in the President's Commission for the Study of Ethical Problems in Medicine and Biomedical and Behavioral Research. Since the Commission can only issue recommendations, it cannot provide the oversight to local human subject committees unless Congress acts or some initiatives are taken by the Department of Health and Human Services which oversees the human subjects regulations. Without some general guidelines establishing ethical norms for human genetic manipulation experiments, the responsibility for developing such policies will be relegated to individual institutional committees. And while the possibilities grow for human genetic therapies, e.g. gene splicing in conjunction with *in vitro* fertilization, and mammalian cloning, free floating public anxieties are in search for an institutional response.

Conclusion: Technology's Double Edge

The medical and commercial applications of gene splicing raise some old problems fashioned in new suits of clothing. The developments in a field bursting with innovative ideas and unlimited potential will put to the test the social guidance systems we presently have. But more so, they will test the moral and scientific wisdom of technologically advanced countries on their capacity to counteract the adverse effects of genetic technology before they are realized and become part of the social and economic infrastructure of society.

Scientific knowledge, said Bacon, is power, and power comes in two denominations, liability and assets. It is unthinkable that nature will release its genetic genie in only a single denomination. That type of technological accounting results in moral bankruptcy. The key to sound technological bookkeeping is in the development of guidance systems whose sole function is to trouble-shoot biotechnology's social impacts. Efforts toward this goal have been notable in the areas of laboratory safety and to a lesser degree, military applications. Initiatives in other areas have been nonexistent.

References

Bacteriological (Biological) and Toxin Weapons. 1975. Convention between the U.S. and other governments at Washington, London, and Moscow, 10 April, 1972. Washington, D.C.: U.S. Government Printing Office.

Budiansky, Stephen. 1982. "U.S. looks to biological weapons." *Nature* 297 (24 June): 615–616.

Chakrabarty, A.M. 1981. *Washington Post* (30 November).

Cooper, T., and J. Fullarton. 1978. "The place of biomedical science in national health policy." Pp. 143–152 in H.H. Fudenberg and V.L. Melnick (eds.), *Biomedical Scientists and Public Policy.* New York: Plenum Press.

Demain, Arnold L., and Nadine A. Solomon. 1981. "Industrial microbiology." *Scientific American* 245 (September): 66–75.

Department of Defense (DOD). 1980. *Annual Report on Chemical Warfare and Biological Research Programs,* 1 October 1979–80 September 1980. (15 December).

Dutton, Diana B., and John L. Hochheimer. 1982. "Institutional biosafety committees and public participation: assessing an experiment." *Nature* 297 (6 May): 11–15.

Fox, Jeffrey L. 1981. "Genetic engineering industry has growing pains." *Chemical Engineering News* (6 April): 17–22.

Goldstein, Richard. 1982. Unpublished letter to Marjorie Sun of *Science,* 7 July.

Hubbard, Ruth. 1977. "Pharmaceutical applications: microbial production of insulin." Pp. 165–169 in National Academy of Sciences, *Research with Recombinant DNA.* Washington, D.C.: N.A.S.

Krimsky, Sheldon. 1982. *Genetic Alchemy: The Social History of the Recombinant DNA Controversy.* Cambridge: The MIT Press.

Lewis, Herman W. 1981. "Role of government in promoting innovation and technology transfer." Pp. 46–53 in *Proceedings, Genetic Engineering International Conference,* 6–10 April. Seattle: Battelle Memorial Institute.

Marshall, Eliot. 1982. "Yellow rain: filling in the gaps." *Science* 217 (2 July): 31–34.

National Academy of Sciences (NAS). 1977. *Research with Recombinant DNA.* Washington, D.C.: N.A.S.

Nossiter, Daniel D. 1982. "Designer genes." *Barrons* 62 (22 February): 8–9, 22.

Office of Technology Assessment (OTA). 1981. *Impacts of Applied Genetics.* April. Washington, D.C.: U.S. Government Printing Office.

Omenn, Gilbert S. 1981. "Government as a broker between private and public institutions in the development of recombinant DNA applications." Pp. 34–45 in *Proceedings, Genetic Engineering International Conference,* 6–10 April. Seattle: Battelle Memorial Institute.

Patterson, W.P. 1981. "Rush to put biotechnology to work." *Industry Week* 210 (7 September): 65–70.

Plasmid Working Group. 1975. *Proposed guidelines and potential biohazards associated with experiments involving genetically altered microorganisms.* 24 February. Institute Archives and Special Collections, MIT.

Preparatory Committee. 1980. *Report for the Review Conference of the Parties to the Convention on the Prohibition of the Development, Production, and Stockpiling of Bacteriological (Biological) and Toxin Weapons and on their Destruction.*

Public Broadcasting Service (PBS). 1981. *Pesticides and Pills: For Export Only.* Television broadcast aired 5 & 7 October.

Rosenfield, Alfred. 1975. *The Second Genesis.* New York: Random House.

Schmeck, Jr., Harold M. 1981. "Patients wait, but is knowledge ripe for human gene therapy?" *New York Times* (31 May).

Schulberg, Francine. 1979. "United States export of products banned for domestic use." *Harvard International Law Journal* 20 (Spring): 331-383.

Science. 1980a "BW and Recombinant DNA." 208 (18 April): 271.

Science. 1980b 209 (12 September): 1282.

Setlow, Jane K. 1979. "How the NIH recombinant DNA molecule committee works in 1979." Pp. 161-163 in Joan Morgan and W.J. Whelan (eds.), *Recombinant DNA and Genetic Experimentation.* Oxford: Pergamon Press.

Shaikh, Rashid, and Michael R. Reich. 1981. "Haphazard policy on hazardous exports." *The Lancet* 2 (3 October): 740-742.

Singer, Maxine. 1979. "Spectacular science and ponderous process" (editorial). *Science* 203 (5 January): 9.

Thornton, Ray. 1981. Statement to the NIH Recombinant DNA Advisory Committee, 10-11 September.

USDHEW-NIH. 1977 Environment Impact Statement on NIH Guidelines for Research Involving Recombinant DNA Molecules, October, in two parts.

Wade Jr., James P. 1981. Memorandum to Assistant Secretaries of the Army, Navy and Air Force, 1 April.

Weir, David, and Mark Shapiro. 1981. *The Circle of Poison.* San Francisco: Institute for Food and Development Policy.

Winner, Langdon. 1977. *Autonomous Technology.* Cambridge, MA: the MIT Press.

Part 2

Technology and the Structure of Agriculture

Chapter 4

Technology and U.S. Agriculture

Peter Dorner

Years after the event, Murray Lincoln, Connecticut's first extension agent, recalled his earliest days on the job:

> Nobody knew what an extension agent was supposed to do. So I asked Dr. Jarvis, who was in the Department of Agriculture, what I should do. He said, "Good Lord, I don't know. Just go down there and find out what the source of income of the farmer is and try to improve it."
>
> One of the first farmers I was told to meet was in Stonington, Connecticut.
>
> As I met this big, raw-boned Yankee farmer, almost without introducing myself, I said to him, "Good Lord, what is your source of income in this area?"
>
> And I have heard this story quoted later, but it actually happened. He looked at me and said, "Source of income, young man? We don't have any around here. We live on lack of expense." (USDA, 1962).

And below is the famous colloquy between one of Steinbeck's Okies and a tractor driver whom he's threatened to shoot for knocking over his buildings and plowing up his land (Steinbeck, 1939:51–52).

> "It's not me. There's nothing I can do. I'll lose my job if I don't do it. And look – suppose you kill me? They'll just hang you, but long before you're hung there'll be another guy on the tractor, and he'll bump the house down. You're not killing the right guy."
>
> "That's so," the tenant said. "Who gave you orders? I'll go after him. He's the one to kill."
>
> "You're wrong. He got his orders from the bank. The bank told him, 'Clear those people out or it's your job.'"

"Well, there's a president of the bank. There's a board of directors. I'll fill up the magazine of the rifle and go into the bank."

The driver said, "Fellow was telling me the bank gets orders from the East. The orders were, 'Make the land show profit or we'll close you up.'"

"But where does it stop? Who can we shoot. . .?"

The agricultural extension service, after the rather inauspicious beginnings described by Lincoln, became one of the key institutions that brought science and technology to the farm. This was part of the same process that gave us the industrial revolution which Steinbeck's Okie tried to shoot, but could not (Griswold, 1948:182). The Okie was really striking out at the failure to devise a public policy to make that revolution serve the entire population. That same scientific and technological process has shaped our elaborately interconnected industrial society and our increasingly interdependent world.

As a people, we often seem obsessed with new techniques—especially mechanical ones. We are enamoured with their invention and development and we have created special institutions to promote their rapid diffusion and adoption. We seem to have shown much less concern for the human impact of some of these mechanical inventions. We look with admiration on the machines and the people who develop them, but we tend to look askance at the people displaced and often impoverished by those same machines. It is, of course, not always an either-or choice: can we not have mechanization along with growing opportunities? can we not have a humane technological society? We are certainly very conscious today of the way in which technologies (especially mechanical technologies) have shaped our socio-economic structure. And we are better able to offer protection and adjustment assistance to people displaced by machines or, like Steinbeck's Okie, caught in the impersonal grip of economic forces they cannot control.

Machine technology has of course had dramatic impacts on the structure of U.S. agriculture. The number of farm people and the number of farms have declined sharply over the past 40–50 years. Simultaneously there was an explosive increase in the average size of farm. Some types of farming, however, were affected earlier and more significantly than others.

What follows is a discussion of some of the key factors associated with the transformation of the socio-economic structures in several areas of U.S. agriculture over the past half century. A few cases are chosen for illustration—cotton because of the dramatic impact of technology on the lives of the people involved, Wisconsin dairy farming in part because of the contrast to cotton farming in terms of the socio-economic structure prevalent before the transformation, and in part because of my own direct experience and research on these issues in this geographic region. The final section presents an evaluation and some judgments about technology and the science upon which it is based as sources for the development of solutions to difficulties—many of which were also brought about by technology.

Technology and Socio-Economic Structures in U.S. Agriculture

The Case of Cotton

One of the most dramatic examples of the impact of technology on socio-economic structures in farming is the case of cotton. Although upland cotton grew well throughout the South two centuries ago, it was a difficult crop to handle; separating the lint from the seeds was a tedious hand-labor process. In 1793, Eli Whitney, a young graduate of Yale University who had accepted a teaching job in South Carolina, invented a practical machine for separating the seeds from the lint. This device, the cotton gin, created new incentives for cotton production which increased from an estimated 10,500 bales in 1793 to 4,486,000 bales in 1861; and "extensive commercial production of cotton led to the expansion of the plantation system, with its use of slave labor. The dependence of the South upon a major export crop, produced largely on slave-operated plantations, set several forces in motion which led to the Civil War" (Rasmussen, 1975: 295-315; James, 1981).

About 150 years after the invention of the cotton gin the introduction of the cotton picker had another profound effect on farming structures. Southern plantation agriculture was transformed after the Civil War, not into a system of small owner-operators, but into a system of sharecroppers (many of them black Americans) who held very insecure tenure and could easily be displaced. The shift from horses and mules to tractors as the major source of power converted many sharecroppers into resident wage laborers or replaced them entirely with off-farm wage workers. A further decline in sharecropping and in overall labor use followed the mechanization of pre-harvest operations and the adoption of chemical weed controls.

All this mechanization did not affect the unskilled labor needed for the cotton harvest. In fact, the demand for seasonal labor per harvested acre rose with increasing yields. The introduction of the mechanical cotton picker in the 1940s, however, virtually eliminated the demand for unskilled labor. The average unskilled labor input per hundredweight of cotton, 33.5 hours in 1940, had dropped to 11.5 hours in 1950 and to 2.4 hours in 1957. At the same time input of skilled labor increased eight fold from 0.3 hours per hundredweight in 1940 to 2.5 hours in 1957 (Day, 1967; see also Street, 1957).

In the first stage of mechanization sharecroppers lost their rights to land but retained a job opportunity (albeit at very low pay) in the cotton harvest. In later stages of mechanization unskilled jobs disappeared and a massive migration of often poorly educated people poured from the South's cotton farms to look for employment in large industrial cities, especially of the North. While such massive displacement of people from farm employment is not the only cause for urban poverty, it has certainly intensified that problem.

The case of cotton is the more significant because both of its key phases of mechanization had major impacts not only on the economic structure of agriculture but also on race relations in the United States. If the South had been settled on a pattern more like that of the North, with family owned and operated farms, the consequences of the cotton gin and later the cotton picker would have been quite different. A community of small property holders controlling its own tax base is able and willing to provide for the schooling of its children and other vital public services. But in the plantation system with its reliance on slave labor, it was the two percent of planters owning large estates with more than 50 slaves who manipulated the levers of political power and social control. Not until the Civil War overthrew the planter aristocracy did the small farmer begin to come into his own (Edwards, 1940:213). Even then, the black sharecropper, and many whites as well, lacked the economic and political base to influence social and economic policy at the local level.

Thus the massive labor displacement in southern agriculture in the 1950s and 1960s involved many poorly educated people unprepared for the jobs becoming available in the growing industries. Farmers and laborers released from a more egalitarian system of family owned and operated farms, as demonstrated in some other regions of the country, might well have been better prepared for urban life.

The more general point here is that the characteristics of the social structure are critical in determining the ultimate effect of technology on the welfare of different groups in society. Institutions must be treated as variables in the development process. Thus in assessing the effects of technology the social-institutional structure must be assessed as well (Dorner, 1971; see also Gotsch, 1972).

By the 1950s and 1960s, relatively high skill levels were required in most employment. Many people pushed and/or pulled out of farming faced a labor market demanding skills they did not have. Subsistence opportunities had virtually vanished. There were far fewer rungs on the ladder of economic opportunity, both in farming and in nonfarming occupations, within reach of those lacking education or special training. Poverty, both rural and urban, would be less acute today if those earlier rural migrants had been better trained and if the agricultural sector had not released so many workers without the skills that contemporary industry demanded. Severe racial and ethnic discrimination intensifed the problems for blacks and other minorities.

Technology and the General Agricultural Transformation

Mechanization and labor displacement, and the farm size expansion that accompanied them, were of course not unique to cotton; these same forces were operating throughout U.S. agriculture after World War II (Agricultural History Center, 1979). A point to remember is that U.S. agriculture developed under conditions of plentiful land and scarce labor. The emphasis, until quite recent years, has been on output and efficiency per person, not per acre. To be sure, there was concern with better livestock breeds, improved soil treatment, more productive plant varieties, etc. Yet the major concern was to extend the capacity of labor through the development of new tools and equipment.

In the U.S., our "green revolution" came only recently. Hybrid corn was first introduced in the late 1920s, but the use of commercial fertilizers did not occur on a massive scale until the 1950s; herbicides and insecticides would not become commonplace for another decade. Sharp yield increases based on biological and chemical technologies are rather new phenomena. These technologies would not by themselves have had a great impact on farm size and socio-economic structure: change in structure was brought about largely by the tractor and related mechanization.

In the United States, as in countries around the world, land-saving technologies — biological and chemical — are essentially neutral to scale or size. This assumes a relatively egalitarian distribution in size of land holdings. A skewed distribution of land holdings and accompanying institutional structures may indeed negate such neutrality and favor the large land holders. The assumption of a relatively egalitarian land distribution pattern is valid for much of the Midwest, but does not correspond to the realities of large parts of the South. Nor is it valid for California where large land grants led to quite a different pattern of settlement.

There is nothing mysterious or complex about scale neutrality and land-saving technologies. Seeds, fertilizers, insecticides, etc. are perfectly divisible inputs usable with equal efficiency on small farms and large farms alike. Water for irrigation may involve some economies of scale, but these economies can be captured by water user associations, cooperatively owned tube-wells, etc. as well as by individual landowners. Likewise, of course, machines can be small or rented, or jointly owned. In this way, machine services too can be made divisible, but the basic factor endowments in U.S. agriculture — with plentiful land and scarce labor — did not encourage these developments.

Mechanization and Farm Size Expansion

In the early years of mechanization, joint ownership of certain machines was indeed quite common. But these did not generally include the tractor or basic tillage tools. Machines used only a few days each year, if timeliness of operation in relation to weather and season was not crucial, were more likely prospects for joint ownership. Joint purchase of a grain thresher by 6 to 10 farmers was quite common, at least in the Midwest, until the late 1940s. Until that time, many farmers stored their bundled, unthreshed grain at the farmstead, in more or less weatherproof stacks or in the lofts over the livestock stables. In either case, the crop was protected and the threat of a long spell of rainy weather was removed. The threshing machine could then be moved from farm to farm with few conflicts over whose grain was to be threshed first and whose last. With the shortage of farm labor brought on by World War II, however, farmers increasingly switched to threshing directly from the field. Timeliness became critical and conflicts arose among the cooperators of a threshing ring. Everyone wanted to be first to avoid losses should the weather turn bad, so most of the shared ownership of machines disappeared by the late 1940s.

During the prosperous years of World War II, farmers had accumulated savings, and credit would soon become more readily available. The machinery companies shifted from war-time production to domestic production, and new and bigger tractors and farm equipment were put on the market. The 1950s saw a major wave of farm mechanization. From 20 to 70 percent of the farms sold, depending upon type of farming, were bought by adjoining farmers to enlarge their operations and achieve the economies of scale associated with the new machines. This process continued throughout the 1960s and most of the 1970s.

Small farmers could survive and stay in business throughout this period — and some did. But they had to settle for less income. They fell behind farmers who were expanding their size of operations and even farther behind people in urban occupations. However, we must differentiate here between a small farmer who got established before 1940 and one who tried to get established in the 1950s. Throughout this period land values and land taxes were increasing at a fairly rapid rate. Those who got into farming in 1940 and paid off their mortgages by 1960 could continue operations at a reduced return. But young families buying farms at the higher land values of 1950 and facing increasingly unfavorable cost/price relations would find it very difficult to meet higher mortgage payments and higher taxes and at the same time provide the income needed by a growing family. The late 1940s was the turning point.

In many cases, the sale of small farms destined for combination with adjoining lands involved older farmers selling out at retirement. In some cases, however, technological developments drove the small farmer out of business or forced him to change his type of farming. The introduction of the mechanical cotton picker and the displacement of Southern sharecroppers, perhaps the foremost example, is not the only one. Another example is from dairying.

On Wisconsin's dairy farms, a major technological innovation of the 1950s made it difficult for small farmers to stay in business. Until then, dairy farmers used 10-gallon cans to store milk in a cooling tank. The cans were picked up by a private trucker each morning and delivered to a processing plant. Beginning in the early 1950s, cans were replaced by refrigerated bulk tanks installed in a special milk-house adjacent to the dairy barn. This was accompanied, or soon followed, by pipeline milking systems pumping milk directly from cow to bulk tank. Now each morning, or in some cases alternate mornings, the milk was picked up by a tank truck. It soon became almost impossible for a dairy farmer to operate without this new equipment, which called for both a major investment and a larger dairy herd than many farmers had at the time. One alternative was to produce milk for delivery to small local cheese factories, but these were also under economic pressures and were being consolidated. New technology created major pressures for farm-size expansion in Wisconsin dairying.

Aside from a few such dramatic cases, which were of course extremely costly and disruptive for the people involved, farmers did have a choice. They could continue without expanding the size of their farms if they were willing to accept deteriorating relative incomes. The only way farmers could keep up with income growth among their nonfarm counterparts was to buy the machines and expand their land base, and this could be done only by combining farms and displacing labor.

Sociological Factors in Farm Size Expansion

Another factor has weighed heavily on the minds of family farmers over the last 30 years. Most every farmer wanted to see the farm kept in the family. Before the 1940s this was usually not a major problem; sons and wives or daughters and husbands who got the home farm considered themselves fortunate. The problem was to figure out how to establish other children in farming, since the land was generally not split up to provide for all; it ordinarily passed to the next generation as a unit. All children usually shared in the will of the parents, but the child lucky enough to become the new operator generally had to buy out the shares of brothers and sisters.

Again, however, things changed after 1940. Farm children were no longer isolated from urban society. Rural electrification gave near-universal access to radio and later television. Most farm children attended high school after 1940, whereas many had not before then. Many farm boys were involved in World War II; after the war jobs in the city were relatively plentiful and young men and women on the farm would not stay home if it meant falling behind in income and sacrificing the amenities urban life seemed to offer. If a farmer did not mechanize and expand, he fell behind in income and his children left the farm for city jobs.

This sort of change in the structure of opportunities is well illustrated by two studies of family farming in Wisconsin. Research in the 1940s documented the relation between the size of the farm business and the life cycle of the farm family (Long and Parsons, 1950). It showed that a Wisconsin dairy farm was a business closely tied to the physical capacity of the farm operator and his family. A young family would build up its business (measured in terms of the number of milking cows) until the farmer reached an age of about 50. Then there were two possibilities. If a son was available to "work his way into the business," the dairy herd was maintained at peak size and the son took over the farm when the father reached the age of 60–65. If no sons or sons-in-law were available, herd size declined after the farmer passed the age of 50, and he would sell out to a new farmer when he reached 60 or 65. The new family would simply start the cycle over again. In the former case, the labor (and strength) supplied by the son came at an appropriate time to offset the declining physical capacity of the father. In the latter case, the waning capacity of the aging father meant a decline in the size of the business.

Similar studies in the 1960s and 1970s show fundamental changes in that cycle. The life-cycle phenomenon and its relation to business size was still evident and pronounced, but the timing and implications had changed. In more recent years, even farmers without sons at home were able to maintain peak herd size until they were about 60 years old. Machine technology had reduced dependence on hard physical labor. Farm wives had become more important in the farm labor force as is pointed out by Haney in chapter 10 in this volume. It is also likely that farm people were healthier and in better physical condition than a generation earlier. Other factors, too, contributed to this greater capacity. Farmers were more knowledgeable about production practices; with their risks reduced, more specialization was possible and secondary enterprises could be eliminated. Increased availability of custom hire of machines was also a factor in some cases. Finally, farmers had achieved coverage under the Social Security system in the 1950s and depended less on their children for care and support in old age. The parents, by the 1960s, had achieved greater independence from their children (Dorner, 1981).

Yet, as noted, farm children had also achieved more independence from their parents. These later Wisconsin research studies show very clearly that if a father and mother were to interest a son or daughter in taking over the farm business, they had to expand operations by the time they reached the approximate age of 50. Even though their own increased capacities would permit them to run the business at peak performance ten years longer than their parents had done, they still had to expand and mechanize in order to provide enough employment and volume of sales to sustain both themselves and a son's or daughter's new family at a constantly rising standard of living.

Again, however, there have been changes in recent years. Part-time farming has become more prominent and it takes a variety of forms. There has actually been an increase in recent years in the number of "small" farms, most of which are operated by part-time farmers. Farm income is generally of secondary importance; income from nonfarm sources is usually larger than that earned from the farm.

A very significant fact is that "part-time" farming has in some cases provided an alternative to farm size expansion. Off-farm employment facilitates the transfer process during the transition period between generations. This may mean that the younger generation will (at least for a time) be part-time farmers and part-time employees in nonfarm activities. Or, it may be that the older generation, father and/or mother, take off-farm employment and work only part-time on the farm. A variety of combinations is possible (Kada, 1980). In any event, the two generations are less dependent on the farm as the sole income source during those critical transition years when both parents and children have physical capacities that cannot be fully utilized and income aspirations that cannot be fulfilled unless the operation is enlarged. Thus, there does seem to be an alternative (also fostered by technology) which was not widely available in earlier years (Dorner, 1981).

Prices and Farm Size Expansion

An additional factor of considerable importance in this transformation process (and also associated with the technological developments of both mechanical and biochemical type) was the changing price structure in agriculture. Throughout the 1940s, farm prices were relatively high. A small farmer got a substantial income boost (certainly in comparison to the depressed prices of the 1930s) from the higher prices even if volume of output remained constant. After the first few years of the 1950s (or more precisely, after the Korean War) farm prices fell. They continued to fall, relative to the prices farmers had to pay for production goods, throughout the 1960s and 1970s. The terms of trade were consistently shifting against farmers after the prosperous 1940s. This shift was temporarily reversed from 1973 to 1975. The short-lived period of relatively favorable price-cost relationships in farming stimulated new investments and raised the hopes of some that export markets would provide a permanent solution to the cost-price squeeze. These hopes were, however, also short-lived.

The only way to maintain the farm family's income from farming was to expand volume of production and to increase efficiency (to lower cost per unit of output). Yet maintaining income was hardly sufficient. Average farm family incomes had always been considerably less than urban family incomes, and urban incomes were rising sharply in these years. Farmers were under pressure from a variety of sources: from the machinery companies introducing and trying to sell new and bigger machines; from a cost-price squeeze; from the prospect of income decline relative to urban workers and other farmers who were mechanizing; and finally, from their own hopes and desires to keep the farm in the family.

Summary

Technology has indeed helped to shape and re-shape the economic and social structure of U.S. agriculture and the nonfarm economy as well. With the decline of the farm population, many small towns and their schools, churches, stores, and social life also declined and even disappeared. But the converse of this relationship—that the particular economic and social structure encouraged the development of particular kinds of technology—must also be kept in mind. The relative scarcity of labor and the pattern of dispersed settlement on family farms in much of U.S. agriculture was itself a key factor in the kind of technology that was developed. The local blacksmith, in close and constant contact with farmers and their needs, was often at the forefront in providing mechanical inventions for the farm. The Land Grant University system, and its experiment stations and extension services, helped to foster the technological revolution in farming. Reliance on migrant workers in some areas for some crops with high seasonal labor requirements also stimulated some mechanization: "The migrant worker question was back of the research leading to the mechanical tomato harvester.

Its rapid adoption was a result of a shortage of experienced migrant workers following the termination of Public Law 78 on December 31, 1964, and the drastic reduction in the importation of foreign labor" (Rasmussen, 1975; see also Rasmussen, 1968).

Had the U.S. developed, for instance, a feudal hierarchy or a communal ownership of land instead of fee simple ownership and family farms; had our social organization developed around the extended family or the tribe rather than the nuclear family living in relative isolation on its farmstead; had our political system fostered central control and management of the economy instead of local autonomy and private enterprise and initiative; our technological developments in agriculture would undoubtedly have been quite different.

Technology and U.S. Agriculture: An Evaluation

Technological developments in U.S. agriculture and the concomitant farm size structure have been the subject of much debate. Critics have attacked the Land Grant Colleges, holding them responsible for these developments (Hightower, 1972). There is much concern about the increasing concentration of farmland in ever fewer production units, non-family corporations in farming, the wide dispersal of pesticides and other chemicals in the environment, the energy-capital-resource intensity of modern agriculture, the huge capital requirements for getting started in farming, the mechanization that continues to displace people and depopulate the countryside, etc. There is indeed much to be critical about.

It would seem entirely consistent with both farm-production efficiency and the frequently stated goal of maintaining the family farm to put restrictions on nonfarm corporations entering farming, to enforce the 160-acre limit for water recipients on publicly funded irrigation projects in the West, to eliminate tax-law advantages that accrue to those with the largest investments, to limit the total amount of government price-support payments a farm can receive, to strike those aspects of credit policies that cheapen the cost of loans for larger borrowers, to pursue vigorous anti-monopoly policies in the farm-supply and food processing segments of agriculture, etc. Simultaneously, assistance to beginning farmers, small- and medium-size producers, farm workers, and rural communities could be strengthened.

While such revisions in policy might help to maintain the current structure and even save the family farm from extinction, they would not reverse, in any fundamental way, 40 years of basic changes in the socio-economic structure of farming. These changes, as noted earlier, are rooted in the machine technology developed throughout those years, in rising labor productivity and declining relative farm product prices, and in the rapid growth in nonfarm family incomes. They also stem from the increasing independence of farm children who agree to "take over" the home farm only if it is big enough to provide incomes comparable to those in nonfarm employment.

Such policy adjustments designed to strengthen the family farm do little or nothing to deal with the other concerns of critics—increased use of chemicals, energy intensity, etc. On these latter points there is a good deal of ambivalence and ambiguity in commonly suggested solutions. In the mid-1970s the world used about 40 million tons of nitrogen fertilizer annually; it is projected that in order to feed the world's population in the year 2000, world agriculture will need to use about 200 million tons each year (Hardy and Havelka, 1975). Such increases, plus concomitant increases in the use of insecticides and herbicides, do present a *serious* threat. We simply do not know enough about the consequent ecological imbalances that may result from such vast growth in the use of chemicals. Does this suggest that we stop using those chemicals? Eventually, maybe so, but we need to develop *through science* some alternative way to supply needed nitrogen and to control pests.

Some suggest that we return to organic farming—a more thorough recycling of organic wastes and green manures, for example—to eliminate the use of chemical fertilizers. It does not automatically follow, however, that yields would remain at present or necessary levels. Rice yields in Japan are the highest in the world. Japan also uses more fertilizer per hectare than any other country. China's rice yields are only a little more than half those of Japan (Wortman and Cummings, 1978); it uses much less commercial fertilizer per hectare than Japan. China has pressed its agricultural output about as far as the use of human power and the recycling of organic matter will permit, and it is now seeking technology to intensify its agriculture in other ways—by producing nitrogen fertilizers domestically and developing its petroleum industry.

What is to be done about the environmental threat? To solve it by abandoning the use of all chemicals means, at least in the short run, a reduced food supply and more hunger in the world. The dual requirements of larger food supplies and lesser dependence on chemicals must be resolved by new directions in research and technological development. Genetic engineering as well as established techniques of plant breeding offer great promise. Research is underway and some progress has already been registered in transferring the legumes' capacity for utilizing atmospheric nitrogen to corn and cereal grains, increasing the efficiency of photosynthesis, domesticating and developing new plant varieties that will grow in saline soils, non-chemical means of insect control, etc.

Similar examples could be cited in regard to energy intensity of the food system. A return to horses and horse drawn equipment is out of the question. The U.S. Department of Agriculture (Gavett, 1975) calculated that a return to the technology and farming practices of 1918 while achieving the production totals of today would require about 61 million horses and mules. We would need about 27 million farm workers, nearly 24 million more than we now use. Feeding those horses and mules would likely take the production from more than 100 million acres of our cropland, a total greater than that now producing crops for export. This change too would increase hunger in the world. Again hope must lie in new directions in science and technology, in moving forward in the development of new sources of energy, energy-saving tillage practices, etc.

The much bigger farm population required in "turning back" to a supposedly simpler and less stressful time would have lower incomes. All personal incomes would be lower, of course, but the drop would be especially pronounced in farming. Many who were born and raised on the farm and who "escaped" from its very hard work and confining life before modern technologies were so widely used can sympathize with an evaluation of all these changes by Paarlberg (1978):

> How should one feel about this? One's reaction is, of course, subjective. As for myself, I feel a mixture of emotions. On the one hand is happiness that the former poverty, isolation, drudgery, and deprivation of farm life is being alleviated. On the other hand is sadness that farmers are being deprived of the sense of unique worthiness they once felt and are becoming an undifferentiated part of a homogenized society. From this throat comes one cheer, maybe two, certainly not three. But what difference does it make, whether there is one cheer, or two or three, or none at all? The change is irreversible. To talk about restoring the uniqueness of agriculture is like talking about putting the chicken back into the egg.

If agriculture were freestanding, isolated from the rest of the economy, we might find some rational or moral principle for "going back" to a simpler technological era. But agriculture is an integral part of the larger economy, with intricate interrelationships and interdependencies, and restructuring of the farm sector can only be undertaken if we are also willing to restructure the industrial sector and the nonfarm economy in general.

Technology has both desired and undesired effects (the latter sometimes unanticipated). It is virtually impossible in our complex, interconnected, interdependent society to make any change that has only "good" (i.e. desired by everyone) consequences. These multiple consequences of new technology are also frequently not all evident in the short run — it may take years before the full impacts are measurable. The guiding principle for the introduction of new technology must be to evaluate the multiple consequences that its introduction will have and to assure that problems (the bads) created are not larger than the benefits (the goods) produced. And these "bads" and "goods" will likely accrue to different groups in the society, depending on the socio-economic-institutional structure. Some "turning back", i.e. rejecting some technologies, rehabilitating others, etc. may indeed be warranted. Specifically with respect to continued growth in farm size, the "technical efficiencies available from the combination of resources for agricultural production can be realized by farms of a relatively modest size" (Penn, 1979). Yet, even some modest "turning back" or holding all farms to the size where their per unit cost of production is lowest would not "restore" the family farm to the status it had in an earlier era. The family farms of today are radically different from those of fifty years ago—so much so that in philosophical, social, and political terms the concept of family farming has lost much of its meaning implicit in that term throughout the nineteenth century.

One need not endorse all the current technology in farming to reach the conclusion that basic technological developments have had a major positive impact on the output of food and consequently on population numbers. The earth's 4 billion people and the many more yet to be born before world population levels off (even with the best of efforts) cannot be fed without continued developments in science and technology. Nor can critical soils and fragile environments be protected and preserved without new scientific knowledge and its well-designed technological application. These must be selective developments, to be sure. All that is new and all that is possible is not desirable. We must by all means give as much public policy attention to alleviating the negative socio-economic and environmental impacts of technological developments as we do to the fostering and the diffusion of new techniques.

References

Agricultural History Center. 1979. "A list of references for the history of agricultural technology." University of California-Davis.

Day, Richard H. 1967. "The economics of technological change and the demise of the sharecropper." *American Economic Review* 57 (June):427–449.

Dorner, Peter. 1971. "Needed redirections in economic analysis for agricultural development policy." *American Journal of Agricultural Economics* 53 (February):8–16.

_____. 1981. "Economic and social changes on Wisconsin family farms (A sample study of Wisconsin farms 1950, 1960, and 1975)." *Research Bulletin* R 3105, College of Agricultural and Life Sciences. University of Wisconsin-Madison.

Edwards, Everett E. 1940. "The first 300 years." Pp. 171-276 in *Farmers in a Changing World. The Yearbook of Agriculture.* Washington, D.C.: Government Printing Office.

Gavett, Earle E. 1975. "Can 1918 farming feed 1975 people." *The Farm Index.* U.S. Department of Agriculture.

Gotsch, Carl H. 1972. "Technical change and the distribution of income in rural areas." *American Journal of Agricultural Economics* 54 (May):326–341.

Griswold, A. Whitney. 1948. *Farming and Democracy.* New Haven: Yale University Press.

Hardy, R.W.F., and U.D. Havelka. 1975. "Nitrogen fixation research: A key to world food?" *Science* 188 (May):633-643.

Hightower, Jim. 1972. *Hard Tomatoes, Hard Times: The Failure of the Land Grant College Complex.* Washington, D.C.: Agribusiness Accountability Project.

James, David. 1981. "The transformation of local state and class structures and the resistance to the Civil Rights Movement in the South." Ph.D. dissertation, University of Wisconsin-Madison.

Kada, Ryohei. 1980. *Part-time Family Farming: Off-Farm Employment and Farm Adjustments in the United States and Japan.* Tokyo, Japan: Center for Academic Publications.

Long, Erven J. and Kenneth H. Parsons. 1950. "How family labor affects Wisconsin farming." *Research Bulletin* 167, College of Agricultural and Life Sciences, University of Wisconsin-Madison.

Paarlberg, Don. 1978. "Agriculture loses its uniqueness." *American Journal of Agricultural Economics* 60 (December): 769-776.

Penn, J.B. 1979. "The structure of agriculture: an overview of the issue." Pp. 2-23 in *Structure Issues of American Agriculture.* U.S. Department of Agriculture, Agricultural Economic Report 438.

Rasmussen, Wayne D. 1975. "The mechanization of American agriculture." Pp. 295-315 in Alan Fusonie and Leila Moran (eds.), *Agricultural Literature: Proud Heritage-Future Promise.* Washington, D.C.: Association of the National Agricultural Library, Inc., and the Graduate School Press, U.S. Department of Agriculture.

_____. 1968. "Advances in American agriculture: the mechanical tomato harvester as a case study." *Technology and Culture* 9 (October): 531-543.

Steinbeck, John. 1939. *The Grapes of Wrath.* New York: The Viking Press.

Street, James H. 1957. *The New Revolution in the Cotton Economy.* Chapel Hill: University of North Carolina Press.

United States Department of Agriculture (USDA), 1962. World Food Forum Proceedings: The Inaugural Event Commemorating the 100th Anniversary of the USDA, 1862-1962.

Wortman, Sterling and Ralph W. Cummings, Jr. 1978. *To Feed this World: The Challenge and the Strategy.* Baltimore: Johns Hopkins University Press.

Chapter 5

Beyond the Family Farm

Frederick H. Buttel

Family Farming: Means or End?

Earl O. Heady, the distinguished agricultural economist, published an article on tenancy and land rental during the height of post-Depression concern with the high frequency of tenancy in U.S. agriculture which has unfortunately and undeservedly remained obscure within contemporary academic circles. Heady (1953) made the argument that academicians and activists, in their singleminded quest to restore owner-occupation of farm land to its pre-Depression levels, had slowly transformed owner-occupation from a means to achieve certain goals or ends into an end in itself. Heady suggested that a substantial degree of tenancy and land rental in an agricultural system was not necessarily undesirable for farmers or for society, given particular institutional arrangements. Indeed, Heady made a cogent case that a certain fraction of land under the ownership of nonfarm entities might play several useful functions in U.S. agriculture, such as enabling young prospective farmers with little capital to enter agriculture without having to tie up their meager assets in expensive farmland; but without leading to the rise of an unproductive class of rentier absentee owners.

An argument similar to Heady's can be made for much of the activism—both academic and nonacademic—focused on family farming issues over the past decade. The virtually universal appeal of family farming has led many to see family farming as an end in itself. In confusing the form and content of social

The author wishes to thank Gene F. Summers and Harriet Friedmann for their helpful comments on a previous draft of this chapter.

relationships in agriculture, activists have been lulled into unquestioned acceptance of many dualities—"corporate" vs "family" farming, "conventional" vs.
"appropriate" technology, "small" vs. "large" farms—that often confuse rather
than illuminate the issues at hand (Buttel, 1981). Most importantly, viewing
family farming as an end rather than as a means to an end causes us to lose sight
of what these ends are. Confusion of the means-ends relationship in agriculture
also prevents us from looking beyond the family farm. To purposefully overstate
the case, if family farming is dead or nearly so, what are viable alternatives to the
family farm which will satisfy, at least in part, the ends for which family farming
was the means to achieve?

Posing such a question forces us to articulate what are the specific ends or
outcomes, real or imagined, that lead us to like family farming. This question
also necessitates inquiry as to whether these ends are being fulfilled. Finally, clari-
fying the means-ends relationship in agriculture should lead agricultural and
rural social scientists to identify plausible alternative means that may be compatible with or a replacement for family farming as a social force in rural America.[1]

Supporters of the family farm no doubt have a variety of ultimate reasons
or "ends" in mind when they lament the continuing problems faced by household production units in U.S. agriculture. Indeed, the nearly universal appeal of
family farming is likely based on the fact that these competing or incompatible
ends are rarely if ever articulated. Nevertheless, for various groups to think
through carefully why they support family farming—that is, to identify the ends
they seek—does not imply that all such ends are equally valid or achievable. For
example, to support family farming in order to prevent large-scale corporate
capital from penetrating the last remaining sphere of decentralized or deconcen-
trated economic activity might be seen as a faulty diagnosis of the problem or an
an unrealistic goal in an advanced capitalist society. To favor family farming
because of a desire to perpetuate the bucolic values that are presumably nurtured in homogeneous rural and agricultural communities might be equally off
the mark.

My own view of the means-ends posture of family farming, while perhaps
excessively personalized and rooted in my own family farming upbringing, is
that the historic forward-looking or progressive rationale for family farming
was *democratization,* and that democratization should remain the benchmark
against which we evaluate the status and prospects of family farming and
inquire into means other than the institution of family farming or household
production in agriculture to achieve desirable ends (see Goodwyn, 1976, for a
parallel view). Family farming historically was a democratic alternative to the
frequently grossly unequal landholding systems of Western Europe and to the
"peculiar institution" of the antebellum Confederacy and its legacy in the share-
cropper/crop lien system of the post-bellum South. Democratization was, of
course, not always served, especially since owner-occupation of distant territories was often a vehicle for the brutal annihilation of native peoples (Mann and

Dickinson, 1980). Nonetheless, the democratic promise of family farming, which, unfortunately, was rendered largely invisible after the demise of Populism (McConnell, 1969; Goodwyn, 1976), was one of ensuring broad access to the productive resources of the society, of narrowing city-countryside disparities of income and wealth, and of ensuring that agricultural development contribute to economic development as a whole, although not on unequal or exploitive terms. That many of these democratic goals for family farming were not fully met should not obscure the fact that these goals remain relevant today and should be the benchmarks against which contemporary family farming and plausible alternatives should be gauged. In particular, we must be attentive to whether family farming is less able than its alternatives to lead to democratization. The remainder of the paper will be devoted to developing an argument about the necessity to look beyond the family farm—to recognize its structural weaknesses, the possibilities other than independent owner-occupation for democratization of agriculture, and political strategies not directly related to preserving the family farm which would otherwise contribute to democratization of or public accountability in agriculture.

The Family Farm in the U.S. Agricultural Structure

Assessing the role of family farms within the contemporary structure of U.S. agriculture became a virtual growth industry during the 1970s. The U.S. Department of Agriculture, in a temporary departure from its historic shallow and defensive posture toward its pro-family farming critics, alone produced three significant monographs on the topic (U.S. Department of Agriculture, 1981; ESCS, 1979; Schertz et al., 1979). Other visible and influential efforts have included those by Rodefeld (1978, 1980) Emerson (1978), Vogeler (1981), Cochrane (1979), Tweeten and Huffman (1980), Shover (1976), Gardner and Pope (1978), Madden and Tischbein Baker (1981), and Stanton (1978).

These numerous empirical assessments of the status of the family farm in U.S. agriculture exhibited several commonalities. All authors emphasized the growing concentration of assets and sales among a relative handful of large farms. Virtually all authors indicated that federal policy played a major role in encouraging or underwriting this concentration process. Other frequent emphases were the growing role of off-farm income and part-time farming, the role of economies of scale, increased farm-level and regional specialization, and the role of technological change. More germane to our purposes is the fact that most assessments of the status of the family farm in the 1970s saw the family farm experiencing significant stress and declining relative to other types of farms. However, not all observers were in agreement on whether the family farm in the late 1970s was still a majority component of U.S. agriculture, or was a minority component which had already undergone a significant demise. For example,

Gardner and Pope (1978), while implying that family farms were declining relative to larger, nonfamily-type farms, were still comfortable with asserting the present and future dominance of family-operated farms. Vogeler (1981) and Rodefeld (1980), on the other hand, concluded that family farms by the late 1970s were a minority sector in American agriculture. The principal factor accounting for these divergent assessments, of course, was the definition used for family farming. Definitions resembling the classical Jeffersonian (and, in a certain sense, Marxist) formulation tended to result in assessments revolving around the minority status of family farms, while utilization of inclusive definitions (historically associated with U.S.D.A. analyses) tended to coincide with pronouncements of the dominance of family farms in U.S. agriculture.

Disparate definitions of farm types were not the only factor leading to imprecision in evaluations of the position of family farming in the 1970s, however. Perhaps the most dramatic and unfortunate commonality among the dozens of such studies was the virtual impossibility of any researcher being able to directly operationalize his or her definition of family farms and other types of farms. The reason underlying these difficulties of operationalization is the inappropriateness of categories in Census of Agriculture data in reflecting socioeconomic—especially noneconomic—characteristics of farms. Further problems included changes in the census definition of a farm and variations in the inclusiveness of enumeration techniques. Nonetheless, the gaps in available public data on the farm sector have further confused an already confusing situation concerning structural change in the farming sector.

To continue the family farm numbers game in the present paper would be tempting but unproductive. Although eschewing a straightforward empirical analysis, I will rely to some degree on recent Census of Agriculture data. The basic thrust of my analysis, however, will be theoretical, both in terms of the role of household production in the agricultural economy and of the political-economic constraints on attenuating the tendency toward the increased prevalence of large nonfamily (or "larger-than-family") farms that is acknowledged by recent analysts regardless of their particular definition of family farming.

Household Production and the Emergence of Agricultural Dualism

Most analyses of structural trends in U.S. agriculture begin with the post-World War II period, since the Depression decade and the War years witnessed the key economic and technical changes which have been accelerated up to the present time. The pre-War and War years saw the rapid adoption of tractor and tractor-related technologies, the development and deployment of modern biochemical farming practices, and the onset of federal farm commodity programs which boosted farm prices and reduced instability in agricultural income. Also, because of the depressed state of the farm economy during the 1930s, which resulted in massive foreclosures of farms, roughly 40 percent of farms in the immediate post-War period were tenant farms. Tenancy was to rapidly decline in the years following World War II, partly because of the institutional milieu of technological change and federal policy. Mechanization, along with the U.S.D.A. decision not to require landlords to distribute government commodity payments to their sharecroppers or tenants, facilitated consolidation of farms and the eviction of tenants. Commodity programs, in placing a floor under farm income, reduced the risks of farming and encouraged larger capital investments. The development of mechanical and biochemical technologies placed a premium on farmers' access to credit, and the favored access of larger farmers to agricultural credit allowed them to assume a superior competitive position in the technological change process. The result in the post-World War II period was a spectacular decline in the number of farms, especially in the numbers of small farms (see, for example, Ball and Heady, 1972).

It is widely recognized that technological change in agriculture – in particular, mechanization – has played a major role in the post-War trends toward fewer and larger farms (see for examples, Cochrane, 1979; LeVeen, 1978; and Dorner, chapter 4 in this volume). It should be kept in mind, however, that the trend toward larger farms was in motion well before the post-War surge of technological change. The rapid introduction of new mechanical and biochemical technologies thus intensified already eshablished trends and did not create a qualitatively different trajectory of socioeconomic change in agriculture.

The predominant thrust of technological change in U.S. agriculture was toward individual farmers' adoption of new inputs that minimized per unit production costs. The early adopters, who usually tended to be large operators with superior credit worthiness, were able to earn temporary windfalls by being the first to utilize cost-minimizing technologies. However, as these cost-minimizing, productivity-increasing technologies became more generalized, they had several important effects. First, utilization of new technology tended to become compulsory for all farmers for them to remain in business. Second, increased aggregate production lowered product price, thereby shifting the benefits of technological change from farmers to the nonfarm sectors.Third, the minimum farm size at which technical economies of scale could be realized tended to increase, placing small- and then medium-sized farmers at an increasing disadvantage relative to their larger counterparts. Finally, increased labor productivity reduced employment opportunities in agriculture and shifted farm returns away from returns to labor and toward returns to capital (Lianos and Paris, 1972). Rapid technological change thus intensified the differentiation of agricultural commodity producers and exacerbated the problems faced by state managers in rationalizing the agricultural sector, a topic that will be discussed shortly.

It is also important to emphasize the rising labor and land productivity, and the corollary tendency toward declining relative returns to agricultural labor, occurred in a context of sharp rural-urban income differentials. Although households reporting farm income still tend to earn lower incomes than other U.S.

households, this disparity was particularly pronounced during the first six decades of this century (see Buttel, 1982a, and the paper by Dorner in this volume). Thus, while many family farmers and agricultural workers were "pushed" off the farm by bankruptcy or technologically-induced unemployment, many farm households were "pulled" off the farm as well by more attractive wage scales in urban places or metropolitan regions.

Most of the post-War period was characterized by familiar dynamics of differentiation (Flinn and Buttel, 1980) and "cannibalism" (Cochrane, 1979). Family farms prospered unevenly and unequally, and hence became more socio-economically differentiated. As a result, smaller and less competitive farms tended to go out of business, and their lands were "cannibalized" by or consolidated into already larger units. At the same time, the character of tenancy and land rental changed dramatically. Small, marginal tenant farms, which were numerically prominent in the 1930s, slowly disappeared under the forces of differentiation and cannibalism. However, despite the rapid disappearance of the marginal tenant or sharecropper farm (especially in the South; Mandle, 1978), the proportion of land rented remained relatively constant. By the 1970s, full- or part-tenant farms tended to be larger and more profitable than full-owner farms (Hottel and Harrington, 1979). The largest of farms increasingly tended to be part-owner farms. These part-owner farms in 1974 averaged roughly 850 acres per farm, compared to 450 acres for full-tenant farms and 250 acres for full-owner farms (Hottel and Harrington, 1979:99).

The most dynamic of farms, part-owner farms, achieved this dynamism by combining land rental to take advantage of economies of scale (both technical and pecuniary) and to reach a favorable cash flow and profitability position, with land ownership. Land ownership, leveraged with profits from rented land and with appreciation of the value of previously-owned land, provided the part-owner with the long-term benefits associated with rising land values. As we will see below, the crucial factor leading to the shifting configuration of land rental and tenancy was the nature of the land market and the tendency for the state to underwrite the profitability of, and hence the market values of and rents from, farm land.

Before discussing the role of the land market and state policy in accounting for the diminishing role of family farms in U.S. agriculture, it is important to note that the 1970s have ushered in a pattern of structural change somewhat different from that of differentiation and cannibalism. This pattern, which I term "dualism," reflects the emergence of a bimodal structure characterized by the increased dominance (in both numbers and sales) of extremely large farm units and by the increased prevalence (in farm numbers) of extremely small farms. A third component of this dualistic transition has been the marginalization and relatively rapid disappearance of medium-sized farms—what many observers have termed the "disappearing middle."[2]

The 1978 Census of Agriculture (U.S. Department of Commerce, 1980) revealed a substantial — and largely unexpected — trend toward increased numbers of farms with sales of less than $20,000 per year. These small or "subfamily" farms increased from roughly 1.514 million in 1974 to 1.585 million in 1978. The increased viability of such subfamily farms has generally been attributed to access to off-farm income and to low debt-asset ratios and flexible cost structures which insulate these small farms from adverse changes in factor and product markets (U.S. Department of Agriculture, 1981; Tweeten and Huffman, 1980; Buttel, 1982c, 1982d).

While small farms were increasing in numbers, relatively large farms — those with sales in excess of $200,000 annually — were rapidly increasing their share of total sales and assets in the farm sector. By 1978, these large farms accounted for only about 3.3 percent of all farms but 44.2 percent of gross agricultural sales (U.S. Department of Agriculture, 1981:46).[3] The 1978 Census of Agriculture at the same time revealed that the most vulnerable sales class category in the farm sector in the late 1970s was that with sales from $20,000 to $39,999, with these farms decreasing from roughly 322,000 in 1974 to 307,000 in 1978. In general, relatively small and relatively large farms were the most dynamic in terms of numerical change, while the medium or middle size categories were the most likely to experience decline. Agricultural economists (e.g., Tweeten and Huffman, 1980) have begun to attribute the "obsolescence" of medium-size family farms to cash flow problems and to the fact that they enjoy neither the advantages of smallness (e.g., ability to acquire significant off-farm income) nor the advantages of bigness (pecuniary economies of scale or the ability to leverage present assets into purchases of additional farm land which earns capital gains in an inflationary land market).

This emergent pattern of dualism implies an unenthusiastic future for the traditional independent, full-time, medium-sized family farm. On one hand, the large farm sector is essentially a nonfamily farming sector. Data for 1974 indicated that well over 90 percent of farms with sales of $200,000 or more reported hired labor expenditures and those large farms with full-time hired laborers averaged roughly eight hired workers per farm (Rodefeld, 1979:53).[4] Thus, the largest of farms, which represent a mere handful of farm units in the country, are basically nonfamily farms, given their dependence on hired labor, and already account for a near majority of farm sales. At the same time, farms with sales of $20,000 or less annually, which would find it impossible to reach the U.S. median family income of roughly $20,000 on farm income alone, accounted for 65.9 percent of all farms and only 8.8 percent of sales (U.S. Department of Agriculture, 1981:43). These small farms can be considered subfamily — and, in a certain sense, nonfamily — farms because their survival is squarely premised on access to significant off-farm income.

These data suggest an interpretation that the traditional independent, full-time, medium-sized family farm is a distinct minority segment of U.S. agriculture and may, in fact, have undergone such a significant demise that it will be impossible to "save" the family farm. Farms with sales from $20,000 to $99,999 annually are conventionally depicted as representing the "middle" or family segment of American farming, and by 1978 these medium-sized farms represented only 27.0 percent of U.S. farms and 34.9 percent of sales. To be sure, these medium-sized farms are still prominent — there were roughly 720,000 of them in 1978 — and, in addition, many farms with sales from $100,000 to $199,999 could be considered family-type operations. However, 1974 data indicated that of farms with sales from $100,000 to $199,999, nearly half reported one or more full-time workers, with these farms reporting full-timeworkers averaging 2.5 workers per farm (Rodefeld, 1979:53) Thus, given available data and foreseeable trends in the direction of dualism, one must be persuaded by two conclusions: First, family farming is no longer dominant in either numbers of or sales from U.S. farms. Second, family farming, while still prevalent in many regions of the country, now represents such a small component of contemporary agriculture that these farms can no longer serve as a base from which to significantly increase the competitive position of family or household units in agricultural production.

Causes of and Limitations of Solutions to the Decline of Commercial Family Farming

The previous section advanced several, essentially ad hoc, explanations for the pattern of structural change in contemporary U.S. agriculture. I will now set forth these explanations in a more systematic way in order to assess the prospects that public policy can readily reverse or significantly attenuate the tendency toward dualism.

One immediate explanation of the emergence of dualism is the tendency toward differentiation among producers in a market economy. Some producers are more efficient or productive than others, leading some farmers to prosper and expand, while others are less dynamic and may be eventually forced to leave agriculture. The tendency toward differentiation, which is inherent in market economies (Friedmann, 1981; Flinn and Buttel, 1980), is a process over which public policy would have little leverage.

Another significant reason for the emergence of agricultural dualism revolves around a set of forces relating to overproduction, public commodity policies, and the land market. Given an agricultural system such as that of the U.S. in which farmers must engage in competition through technological change and reduction of per unit production costs (LeVeen, 1978; Cochrane, 1979), there is a tendency toward increasing productivity and, hence, overproduction. Overproduction has traditionally been dealt with through government commodity programs which place a floor under farm product prices and farm income. Many observers have argued that this floor phenomenon, which is dictated politically by the need for elected politicians to attract the still substantial farm and rural vote, leads to concentration in the farm sector by reducing insecurity of investment and unequally distributing program benefits to larger producerAs noted earlier, reduction of the insecurity of investing in farming makes agriculture attractive to larger investments (Spitze, *et al.,* 1980). Also, program payments have tended to be distributed unequally in the favor of larger producers (Schultze, 1971).

The risk reduction and regressive distributional consequences of farm programs to address endemic overproduction problems may be subordinate, however, to the consequences of these programs for the agricultural land market. Since agriculture continues to have a dispersed ownership structure, and because no one farmer can control the conditions of competition, farmers receiving commodity program benefits will tend to bid these revenues into the price of land on local land markets. Benefits from farm programs (and also, in many instances, from technological change) will thus tend to be capitalized in asset — especially land — values.

Resultant land price inflation has several important consequences for the family farm and for structural change in agriculture. First, and most importantly, rising land prices greatly increase barriers to entry, reducing the ability of prospective young family farmers from entering the agricultural sector. Second, land price inflation underwritten by government policy has a major, but often invisible, distributional impact within the farm sector. The principal beneficiaries of land price inflation tend to be large farmers and landowners for whom land speculation is a major, if not overriding, reason for the purchase of land. Rising land prices result in windfall benefits for landlords in the form of growing rent receipts, while large nonfamily farmers turn asset appreciation into credit worthiness for the acquisition of still more farm land. Smaller operators, especially those whose purpose for owning farm land is to engage in farming on an indefinite basis, will benefit far less from land speculation. These farmers will be faced with rising rents, taxes, and interest payments. Most importantly, however, they must leave the agricultural sector in order to realize capital gains from land inflation, creating the strong possibility that their farm will be purchased by an already large land owner with superior credit worthiness. Therefore, land inflation tends to result in a rigidification of landholding structures due to increased barriers to entry and to exacerbation of the tendency toward differentiation among farmers.

A final factor leading to dualism in agriculture is the role of the systems of taxation characteristic of the U.S. and other advanced industrial societies. These societies are characterized by tax systems which subsidize capital investment and expansion through provisions such as accelerated depreciation allowances and investment tax credits. These tax provisions, which are *not* a specifically *agricultural* policy but instead apply to all income earners, have played a major role in underwriting concentration of assets and sales in U.S. agriculture (Raup, 1978;

Spitze, *et al.,* 1980). By subsidizing investment and expansion, such tax savings represent a transfer of resources from taxpayers to expanding farmers through this underwriting of purchases of land and machinery. These tax provisions have a differentiating impact on farm structure because farmers are not equal in their credit worthiness. Farmers with high credit worthiness, who usually are large operators or nonfarm investors (Spitze, *et al.,* 1980), are thus the principal beneficiary of these tax benefits.

Against this background of a multiplicity of forces leading to concentration of sales and assets in agriculture, is it plausible to suggest that public policy can attenuate or reverse these forces? My argument is that such policies are essentially infeasible for both political and economic reasons.

One major ground for skepticism about political strategies to restore the family farm to its immediate post-World War II status is that differentiation, which is inherent in capitalist-market economics, cannot be fully attentuated by foreseeable political means. As long as there remains a market economy in agriculture, the initially unequal asset positions of farmers will be further magnified by their unequal productive and management capacities. While state policy clearly has tended to exacerbate the differentiating tendencies caused by factor and product markets, elimination of the biases of these policies will not be sufficient to end, let alone reverse, the differentiation process.

A second rationale for arguing the political-economic infeasibility of public strategies to save the family farm is that many of the public policies that exacerbate differentiation tend to be those which transfer resources from the nonfarm sector to the farm sector. As Tweeten and Huffman (1980) have noted, farmers, regardless of their size-class situation, tend to be opposed to the cessation of policies such as price supports and tax subsidies that otherwise transfer resources from nonfarmers to farmers. Thus, it would be quite unlikely that farmers would favor the elimination of commodity programs or modification of tax policies which subsidize capital investment and expansion in agriculture.

Third, for public policy to reverse the historic process of differentiation, this policy would require that overt penalties to scale be built into each major public program. For example, commodity programs would have to provide substantially larger benefits to small and medium-size farmers than to larger operators. Tax policy would need to be changed to discourage expansion. Credit policy would require a shift toward ineligibility of large operators for access to publicly provided, subsidized, or administered credit. Overt penalties to scale appear to be extremely unlikely. Not only would favoritism toward small units be bureaucratically problematic, but powerful interest groups in agriculture — especially the American Farm Bureau Federation and major commodity organizations — would bitterly oppose and be able to prevent the establishment of such policies.

A fourth factor mitigating against the feasibility of public policy for saving the family farm is the formidable barriers faced by the federal government in rationalizing overproduction and resultant farm income instability problems in the context of a private market in land. Since the invocation of federal commodity programs, farmers have come to expect that the state has a responsibility to maintain farm income and, in effect, to underwrite costs incurred in inflationary land markets. State officials will thus be faced with strong pressures to maintain farm prices, but will be unable to prevent the partial capitalization of these benefits into land values and the socioeconomic differentiation that tends to result from capitalization process (Breimyer, 1977).

A final political-economic barrier to the development of agricultural policies for enhancing the status of the family farm revolves around the circumscribed political capacities of a social class — independent commodity producers or small property owners — that tends to be economically subordinate but, at the same time, propertied. Many analysts (e.g., Vogeler, 1981) have been unduly optimistic about the potentials of small and family farmers to mobilize themselves to seek favorable policy changes. As noted earlier, many such policy changes would curb the prerogatives of farmers *qua* property owners. These policies hence would tend to be resisted by family or sub-family farmers who one would otherwise expect to be the beneficiaries (relative to larger operators). The ambiguous class position of family farmers thus at least partially explains why the major groups representing farmers (especially the Farm Bureau) have tended to operate primarily in terms of defending existing property relations within agriculture and between agriculture and industry. The otherwise progressive impulse of the Farmers Union has also been blunted by the tendency for its members' perceived interests as property owners to retain primacy over the social reformist values and motivations upon which the organization was founded. The crucial point is that while family farmers are by no means inherently politically reactionary, neither are they inherently progressive. It would thus appear that a progressive-reformist sentiment — one that sacrifices certain prerogatives of capital for economic security and other collective goals — is not likely to emerge and be sustained even among the ranks of those family farmers who continue to be marginalized within the agricultural political economy.

Beyond the Family Farm: Theoretical and Policy Implications

The foregoing has set forth three major postulates which together imply an unenthusiastic future for the traditional family farm. First, medium-sized, full-time family farms — farms owned, managed, and operated primarily by family members and capable of sustaining the reproduction of the farm business and farm family with agricultural income — are now a minority segment of U.S. agriculture in terms of farm numbers and sales. Second, a very powerful set of forces has led toward differentiation and dualism. Third, there are major — if not intractable — political-economic barriers to the development of policies to counteract these differentiating forces and to maintain or enhance the position of family farms. These three conclusions combine to suggest a continued trend toward the "disappearing middle" and "family farm obsolescence" (Tweeten and Huffman, 1980:77).

Acceptance of this conclusion, which is be no means heretical and which has long been nurtured silently among many agricultural economists (see, for example, Knutson, *et al.,* 1980), does not lead to a singular policy implication. The possible policy implications must be evaluated in terms of the ends for which the family farm has ostensibly been the means. Thus, several potential implications of the demise of the family farm – that the trend is desirable because large farms are most efficient (Knutson, *et al.,* 1980), that the government can best help farm families by providing more nonfarm jobs in rural areas (Paarlberg, 1980:199), and so forth – must be judged in terms of an explicit set of socioeconomic goals for the agricultural sector. It was suggested earlier that democratization has been the fulcrum of progressive sentiment toward family farming. The remainder of this chapter will be devoted to how alternatives to the traditional family farm can help move toward democratization of U.S. agriculture.

Before discussing these alternatives, it is important to address one important theoretical implication of the decline of commercial family farming which has been a longstanding postulate in influential portions of the Marxist literature. This postulate concerns the technical superiority of capitalist over noncapitalist forms of production in agriculture and the inevitable tendency for agriculture to experience the predominance of capital-labor relations typical of nonfarm industry. This line of argumentation, the initial codification of which has generally been attributed to Kautsky (1899), has typically been employed to suggest that the only long-term progressive force in agriculture is the agricultural proletariat. Thus, in this view, progressive political strategies should straightforwardly seek to advance the interests of agrarian wage laborers against the interests of agrarian capital, rather than to court ultimately reactionary property-owning peasants or family farmers.

The argument is not so much wrong as it is temporally indeterminant (Hussain and Tribe, 1981a, 1981b); that is, full-time wage labor and the capital-labor relation are clearly growing in importance in U.S. agriculture, but these transitions are very gradual and do not conform organizationally, ideologically, or spatially to those which have characterized nonfarm industry (Newby, 1978 and chapter 6 in this volume). It is essential to recognize the barriers to capitalist penetration of agriculture (Mann and Dickinson, 1978; Mann, 1982) and that U.S. agriculture is, in the main, dominated by petty capitalist (or "larger-than-family") farmers who employ a dispersed wage labor force with relatively few laborers on each farm. Full-time agricultural wage laborers typically exhibit many noneconomic ties with their employers – for example, growing up in the same rural community, partial in-kind remuneration (especially in the case of housing), and so forth. Because of these noneconomic or nonmonetary bonds, full-time farm workers tend to be subject to paternalistic ideological control

(Newby, 1978) and are very hesitant to mobilize collectively against their employers (Wilkening, 1972).

The key implication of my criticism of viewing the large, nonfamily sector of U.S. agriculture in terms of the capital-labor relations that characterize non-farm industry is *not* that the agricultural labor markets are unimportant arenas for the democratization of American agriculture. Indeed, I will suggest below that agricultural labor considerations will be extremely important in developing appropriate strategies. Nevertheless, the point remains that the dominant issues in agricultural politics are unlikely to be reducible to those of capital-labor relations and that hired farm workers generally do not possess the political capacity to be a vanguard of a movement for the democratization of agriculture.

The focus of this chapter has been on developing theoretical and policy alternatives to the family farm as the organizing force of efforts to improve the quality of agricultural life in the U.S. It is therefore useful to specify the notions to which I will pose alternatives. The received strategies generally fall into three categories: (1) advocating public policies which give preference to small over larger farms, (2) the enactment of laws to ban corporate agriculture and to prevent corporate farmers from "unfairly outcompeting" family farmers, and (3) provoking change in public research and extension institutions so that these institutions develop technologies and programs serving family rather than corporate farmers. Elsewhere (Buttel, 1981, 1982b) I have argued that these strategies have a number of limitations. They are unrealistic about the prospects for enabling small farmers to outcompete larger farmers. These strategies also are often based on a "family-corporate" farmer dichotomy that fails to recognize that most large farmers are not corporate-industrial farms in the conventional sense, but rather are large family-owned and -managed businesses. Third, these traditional strategies have often reflected a naive form of technological determinism which suggests that the development of "small farmer technologies" will enable modest-size operations to be at a competitive advantage *vis-a-vis* larger-than-family operations; dubious corollaries—that small is beautiful and large is evil, and that modern technological forms such as large tractors are inherently "inappropriate technology"—have also emerged.

When I argue for looking "beyond the family farm," I do not advocate a view that agricultural production focused around household or family labor is either archaic or implausible. What I do argue for, however, is the notion that it is *no longer possible, or perhaps even desirable, to restore the traditional, independent family farm to the position it occupied at the end of World War II.* My view is that a system of *atomistic, independent* family farms in a context of contemporary market institutions is not a viable means to sustain the position of household ownership and labor in farming.

While there are no doubt a variety of plausible ways of looking beyond the family farm, I will premise my argument on the notion that the key theoretical and policy challenge is to both encourage and force current larger-than-family farms to rely more on family labor and less on wage labor. This formulation, it might be noted, is in contrast to the traditional one advocating the "deconcentration" or "dedifferentiation" of large or corporate farms—that is, utilizing political means to cause the liquidation and ultimate fragmentation of large farming units. This latter route may well be successful in certain contexts, but it flies in the face of ominously powerful trends toward economies of scale and the other financial benefits of operating large farms.

Three potentially promising strategies for re-establishing the primacy of household labor in U.S. agriculture will be discussed below.[5] They include: (1) the development of "group" or "syndicate" farms, (2) strategies to improve the condition of agricultural wage laborers on larger-than-family farms and to discourage the expansion of this wage labor segment, and (3) local strategies to publicly regulate the land market so as to reduce the speculative process in land markets.

The notion of group or syndicate farming as a means to combine large-scale farming and family labor has not been widely discussed in the U.S., although a number of farms of this type exist in Canada, France, Australia and other industrial societies. Group farming is generally taken to mean the pooling of certain farm assets—almost always machinery and labor, but in many instances land and other capital items—to achieve scale economies, to conserve scarce capital resources, to develop a multifamily division of labor, or to achieve other goals. My involvement in research on group farming in Saskatchewan (Gertler and Buttel, 1981; Gertler, et al., 1981) leads me to a conclusion that group farming is a promising mechanism for the accommodation of large-scale agricultural production with the capital and labor resources of families. Group farms achieve significant economies by comparison with their independent family farmer neighbors, especially in terms of lower per unit machinery costs and the ability to diversify their enterprises within the constraints of scarce capital and family labor resources. The Saskatchewan case study of group farming also suggests ecological and social advantages (e.g., greater flexibility with regard to taking vacations) relative to the traditional independent family farm.

Group farming has generally been conceived of as an option best suited for independent family farmers who are experiencing economic and other problems. Empirically, most group farms in advanced industrial societies have tended to be formed by several formerly independent farm families pooling their capital and labor resources (Gertler and Buttel, 1981). However, group farming might well be equally attractive to already large farmers who depend heavily on hired labor. For example, a large farmer might enter into a group arrangement with another comparable farmer, enabling the two farmers to jointly purchase larger equipment which eliminates the need for part-time or full-time hired labor. Another alternative might be for a large farmer to forge a group farming arrangement with a few smaller family farmers, with the large farmer contributing disproportionate capital resources in exchange for the disproportionate labor contributions among the families with smaller amounts of land and other capital items.

The group farming arrangements discussed above have received little serious consideration among the U.S. agricultural research community, despite their potential attractiveness. While group farming is not without significant problems (see Gertler and Buttel, 1981), the promise of group farming in adapting the labor and capital of households to the exigencies of large-scale farming suggests that much more theoretical and practical attention should be paid to this organizational alternative. The group farming alternative may be especially promising if agricultural labor issues are approached in ways that take more seriously the role of full-time hired labor in American farming.

I noted earlier that a significant segment of the agricultural labor force consists of a dispersion of full-time hired workers across a number of larger-than-family farms. While agricultural labor issues in the industrial farming settings of California, the Southwest, and Florida have received considerable and deserved attention (see, for example, Friedland, et al., 1981), the full-time agricultural labor force in nonindustrial farming settings has been almost totally ignored (Newby, 1980). This nonindustrial, full-time agricultural labor force increasingly can be considered the backbone of nonfamily farming in the U.S. Moreover, the stability of larger-than-family farming is delicately premised on access to cheap hired labor (as has been admitted by Knutson, et al. [1980], who are among the more overt proponents of this form of agriculture). At the same time, this agricultural labor segment is very difficult to organize due to its dispersion and to the paternalistic forms of social control that predominate. These factors, plus the fact that agricultural labor issues are politically sensitive among farmers in general, have undoubtedly led to a lack of attention to this important labor market.

The difficulties of conducting research on and organizing full-time hired workers in nonindustrial agriculture necessarily dictate strategies different from those employed in California, Texas, and Florida. In particular, it will largely fall upon nonfarm groups to document the unfavorable economic conditions of these workers and to advocate in federal and state governmental arenas that all forms of protective labor legislation—especially concerning minimum wages, nonwage benefits, and payment for overtime—be fully extended to agriculture. These strategies will no doubt be conflictual and highly controversial. However, they are also an important key to attenuating the expansion of larger-than-family as well as corporate-industrial farms. In effect, the availability of cheap hired labor becomes bid into land values and provides barriers to entry for prospective farmers with fewer assets. Indirectly, cheap hired labor competes with family labor, exaggerating the scale economies that are available to expanding farmers and reducing the magnitude of the principal advantage—the cost flexibility due to the lack of a wage bill for hired workers—that family farmers have over nonfamily farmers (Friedmann, 1981).

The third component of this agenda for looking beyond the family farm — public regulation of land markets — has several conceptual similarities to current strategies that have already been implemented or that are presently being discussed. Preferential taxation of farm land to curb land speculation at the urban fringe has roots back to the 1960s. More interventionist programs such as the purchase of development rights have briefer histories and fewer successes. Several states have legislation banning or restricting corporate purchases of agricultural property. These programs, although laudible in many respects, have, however, assumed that the speculative impulse in agricultural land markets can be traced to land price pressure at the urban fringe and resultant high taxes, or to nonfarm corporate involvement in local and markets. While not denying the importance of these forces, existing programs neglect the role of federal policy in underwriting land price inflation in remote agricultural regions or in states where corporations are highly restricted in their ability to purchase farm land. Programs such as preferential taxation thus not only fail to address the broader origins of agricultural land inflation, but may, because of the effective transfer of resources from nonfarmers to farmers, further exacerbate the inflationary problem.

National legislation for public regulation of land markets as has been implemented in France, Sweden, and portions of Canada would have greater leverage on the land market problem and would be more attractive than many current land policies. However, national legislation would appear to be politically unrealistic given the already large community of interest behind unregulated, inflationary land markets (Buttel, 1982b). Nevertheless, there have been several promising localized programs in the U.S. that are capable of playing comparable roles.

One such example is the Minnesota Farm Security legislation of 1976 (Laws of Minnesota, 1976, Chapter 210) which is focused around a publicly subsidized loan program. The legislation created a seven-member advisory board to screen applicants for farm land purchases and gives preference to credit allocations to entering farmers and to applicants for intergenerational transfers of farm property. A more localized institution for intervention in the land market is the community land trust (CLT). CLTs are usually organized as nonprofit corporations and are governed by boards of trustees elected from leaseholders of trust-owned lands and from the community at large. CLTs function by purchasing land and retaining the "development" and "transfer" rights to this land, with the users of the land typically retaining partial ownership rights such as ownership of buildings and equity in farm improvements. The CLT leases land to farmers on a long-term, secure basis. Most arrangements involve a lifetime or 99-year renewable lease which is transferable to heirs. The underlying rationale of land trusts thus is to enable the leaseholders to be insulated from the broader national pattern of land inflation and to introduce community interests and input into the use of land.

It should be noted that the Minnesota Farm Security legislation and community land trusts are very imperfect instruments for effective intervention in the land market. The Minnesota legislation may encourage farmers to assume a higher level of debt than may be desirable for the farmer or for society, while CLTs are obviously limited in their ability to assemble the financing necessary for expanding the amount of land under their control. Nevertheless, they are positive strategies to bring some degree of public control to land market institutions and may be taken as illustrative directions toward which future public policies might evolve.

The Challenge for Rural Sociology

I would like to conclude this paper by making the observation that the transparently political comments in the previous pages should be construed as being more than a normative agenda for agricultural change. My intention has been to set forth some key conceptual and structural issues with which my readers might agree as being important regardless of their views on the future politics of agriculture. The configuration of recent changes in U.S. agriculture provides a strong challenge to a wide variety of influential social and economic theories. The strong "larger-than-family" farm character of contemporary farming thus far has been handled awkwardly by theorists who posit the primacy of "independent commodity production," "capitalist-industrial farming" or other "pure" forms of agricultural organization. The shifting and often internally contradictory role of state intervention in agriculture cannot be neatly explained by widely accepted theories of the state in advanced industrial societies. The organization and dynamics of American agriculture will thus continue to present a fertile field of inquiry for rural and not-so-rural sociologists.

Many readers will no doubt find fault with the implications I have advanced for research and practice in thinking beyond the family farm. However, I hope to have been persuasive in making several key arguments: (1) that the decline of the family farm raises key challenges for a wide spectrum of social scientists and policy makers, (2) that meaningful policies to address the family farming question must be based on a clearer sense of the ends for which family farming purportedly is to be the means, and (3) that the grounds for specifying these ends will be a combination of the normative and empirical. Rural sociologists must play an important role in addressing each of these issues. To accomplish these tasks will require styles of inquiry — multidisciplinary and comparative — that have not been traditional strengths of U.S. rural sociology. While not sacrificing our identity as rural sociologists or sociologists, the emergent problems that the discipline will face dictate a multidisciplinary agility which must, at a minimum, include greater knowledge of the economics of agriculture. Also, understanding the origins and roles of household production in agriculture in other Western societies will help rural sociologists to better understand agriculture in our own society. Finally, this work may have a very positive influence on the larger discipline by providing some basic knowledge on groups such as independent business people and small property owners that have been underresearched in mainstream sociology. In sum, we face exciting challenges which can only be met by the kind of intellectual courage which is being honored in this Festschrift.

Notes

1. At some risk of deliberately flaunting the volume editor's proscriptions against "eulogizing" Professor Wilkening, it is important to note the origins of the thoughts presented herein in a paper written by Wilkening (1972) a decade ago. Wilkening wrote:

   The concept of the family farm has implicit within it several criteria stemming from the agrarian traditions of Europe and from early American philosophy. But the attempt to define what these are becomes more difficult with increased specialization and differentiation of the farm enterprise. A reformulation is needed which can provide the basis for evaluating policies and programs. Perhaps family ownership of the farm as such is not as important as the principle that every family engaged in agricultural production should have the right to use land for their subsistence and residence. This would provide both stability as well as security for those engaged in agriculture. An alternative would be provision for sharing in the ownership of land through a cooperative or a corporation, but in either case with a limit on the amount of land owned by a single person, family or group. Perhaps the basis for such a limit should be determined by decreasing benefits or by imposing absolute size limits (1972:15).

2. It should be noted that the transition toward dualism is just that — an empirical trend, rather than a completed process. In particular, the "middle" of full-time, medium-sized, independent family farms has by no means fully disappeared, although their absolute and relative numbers are in decline.

   Also, it is crucial to emphasize that aggregate statistical trends disguise major differences between various agricultural commodity systems. For example, tomatoes and lettuce are generally produced on large-scale industrial farms, while tobacco and cranberries tend to be produced on smaller farms. The recent book by Friedland, *et al.,* (1981) provides an exemplary case study of the lettuce commodity system while making a more general case for a commodity systems approach to the sociology of agriculture.

3. A matter that will be taken up later concerns whether large nonfamily farms consist primarily of capitalist or corporate farms (in which the manager is not the owner of the farm property) or "larger-than-family" farms (in which the manager is the major or a significant owner of the farm assets). I have argued elsewhere (Buttel, 1981) that larger-than-family farms are the dominant segment among large nonfamily farms in the U.S. based on both numbers of farms and total sales.

4. The 1978 Census of Agriculture revealed an intensification of a decade-long trend toward greater hired labor inputs in U.S. agriculture. Among farms with sales of $2,500 or more, the number of hired workers increased from 712,715 in 1974 to 953,323 in 1978 (U.S. Department of Commerce, 1980). The increase in the number of worker-days of hired labor has been primarily accounted for by growth in full-time wage work (by workers employed 150 or more days per year), while temporary wage

work (by workers employed less than 75 days per year) has exhibited a slight absolute decline in total worker-days. Most importantly, the numbers of full-time farm workers and total worker-days have been increasingly concentrated, along with cash receipts and land operated, on fewer and larger farms (Rodefeld, 1979).

Two other comments on hired labor in agriculture are relevant to the discussion that follows. First, purely quantitative indicators of the extent of hired labor on farms (e.g., the U.S.D.A. criterion of less than 1.5 man-years of hired labor implying a family operation) may be a poor substitute for determining the extent to which a farm operation is dependent on wage labor. Given the inflexible cost structures of most contemporary large, nonfamily farms, most would not have the option of greatly decreasing the intensity of the farm operation in order to reduce their dependence on hired labor. Second, the social relations associated with wage labor may change in ways that are impossible to gauge with shifts in the man-hours of hired labor. In particular, several decades ago many hired farm workers were the sons of neighboring farmers who entered into farm wage work to gain farming experience and accumulate capital for the eventual transition into an independent farming career. Mobility from farm laborer to independent farmer has no doubt been almost totally eliminated. This shift, one would expect, has had significant impacts on the social relations between farmer-employers and their hired workers.

5. Much of this section is taken from Buttel (1982b).

References

Ball, A. Gordon, and Earl O. Heady. 1972. "Trends in farm and enterprise size and scale." Pp. 40-58 in A.G. Ball and E.O. Heady (eds.), *Size, Structure, and Future of Farms*. Ames: Iowa State University Press.

Breimyer, Harold F. 1977. *Farm Policy*. Ames: Iowa State University Press.

Buttel, Frederick H. 1981. "American agriculture and rural America: challenges for progressive politics." Ithaca, N.Y.: *Cornell Rural Sociology Bulletin* No. 120, March.

_____. 1982a. "Farm structure and rural development." In W.P. Browne and D.F. Hadwiger (eds.), *Rural Policy Problems*. Lexington, Mass.: Lexington Books.

_____. 1982b. "Agricultural land reform: issues and prospects." In C.C. Geisler and F.R. Popper (eds.), *Land Reform, American Style*. Montclair, N.J.: Allanheld, Osmun & Co.,

_____. 1982c. "The political economy of part-time farming." *GeoJournal* 6.

_____. 1982d. "The political economy of agriculture in advanced industrial societies: some observations on theory and method." In S.G. McNall and G.N. Howe (eds.), *Current Perspectives in Social Theory*. Greenwich, Conn.: JAI Press.

Cochrane, Willard W. 1979. *The Development of American Agriculture*. Minneapolis: University of Minnesota Press.

Emerson, Peter M. 1978. *Public Policy and the Changing Structure of American Agriculture*. Washington, D.C.: U.S. Congressional Budget Office.

Economics, Statistics, and Cooperative Services (ESCS). 1979. *Structure Issues of American Agriculture*. Washington, D.C.: U.S. Department of Agriculture.

Flinn, William L., and Frederick H. Buttel. 1980. "Sociological aspects of farm size: ideological and social consequences of scale in agriculture." *American Journal of Agricultural Economics* 62 (December): 946-953.

Friedland, William H., Amy E. Barton, and Robert J. Thomas. 1981. *Manufacturing Green Gold.* Cambridge: Cambridge University Press.

Friedmann, Harriet. 1981. "The family farm in advanced capitalism: an outline of a theory of simple commodity production in agriculture." Paper presented at the annual meeting of the American Sociological Assocation, Toronto, August.

Gardner, B. Delworth, and Rulon D. Pope. 1978. "How is scale and structure determined in agriculture?" *American Journal of Agricultural Economics* 60(May): 295-302.

Gertler, Michael E. and Frederick H. Buttel. 1981. "Property, community, and resource management: exploring the prospects for group farming in North America." Paper presented at the annual meeting of the Rural Sociological Society, Gwelph, Ontario, August.

Gertler, Michael E., A.F. MacKenzie, and Frederick H. Buttel. 1981. "Resource management implications of group farming in Saskatchewan." Paper presented at the annual meeting of the American Association for the Advancement of Science, Toronto, January.

Goodwyn, Lawrence. 1976. *Democratic Promise.* New York: Oxford University Press.

Heady, Earl O. 1953. "Fundamentals of resource ownership policy." *Land Economics* 29(May):44-56.

Hottel, Bruce, and David H. Harrington. 1979. "Tenure and equity influences on the incomes of farmers." Pp. 97-107 in ESCS, *Structure Issues of American Agriculture.* Washington, D.C.: U.S. Department of Agriculture.

Hussain, Athar, and Keith Tribe. 1981a. *Marxism and the Agrarian Question, Volume I: German Social Democracy and the Peasantry, 1890-1907.* Atlantic Highlands, N.J.: Humanities Press.

_____. 1981b. *Marxism and the Agrarian Question, Volume II: Russian Marxism and the Peasantry, 1861-1930.* Atlantic Highlands, N.J.: Humanities Press.

Kautsky, Karl. 1899. *Die Agrarfrage.* Stuttgart: Dietz.

Knutson, Ronald D., Ed Smith, James W. Richardson, and Christina Shirley. 1980. "Maximizing efficiency in agriculture." Pp. 115-122 in *Increasing Understanding of Public Problems and Policies.* Oak Brook, IL: Farm Foundation.

LeVeen, E. Phillip. 1978. "The prospects for small-scale farming in an industrial society: a critical analysis of *Small is Beautiful.*" Pp. 106-125 in R.C. Dorf and Y.L. Hunter (eds.), *Appropriate Visions.* San Francisco: Boyd and Fraser.

Lianos, Theodore P., and Quirno Paris. 1972. "American agriculture and the prophecy of increasing misery." *American Journal of Agricultural Economics* 54(November): 570-577.

Madden, J. Patrick, and Heather Tischbein Baker. 1981. *An Agenda for Small Farms Research.* Washington, D.C.: National Rural Center.

Mandle, Jay R. 1978. *The Roots of Black Poverty.* Durham, N.C.: Duke University Press.

Mann, Susan A. 1982. "Obstacles to the capitalist development of agriculture: an analysis of the class structure and uneven development of American agriculture, 1870-1930." Unpublished Ph.D. Dissertation, Department of Sociology, University of Toronto.

_____. and James M. Dickinson. 1978. "Obstacles to the development of a capitalist agriculture." *Journal of Peasant Studies* 5(July):466–481.

_____. and James M. Dickinson. 1980. "State and agriculture in two eras of American capitalism." Pp. 283–325 in F.H. Buttel and H. Newby (eds.), *The Rural Sociology of the Advanced Societies*. Montclair, N.J.: Allanheld, Osmun & Co.

✓ McConnell, Grant. 1969. *The Decline of Agrarian Democracy*. New York: Atheneum.

Newby, Howard. 1978. "The rural sociology of advanced capitalist societies." Pp. 3–30 in H. Newby (ed.), *International Perspectives in Rural Sociology*. Chichester: Wiley.

_____. 1980. "Rural sociology—a trend report." *Current Sociology* 28 (Spring): 1–141.

Paarlberg, Don. 1980. *Farm and Food Policy*. Lincoln: University of Nebraska Press.

Raup, Philip M. 1978. "Some questions of value and scale in agriculture." *American Journal of Agricultural Economics* 60(December):303–308.

Rodefeld, Richard D. 1978. "Trends in U.S. farm organizational structure and type." Pp. 158–177 in R.D. Rodefeld, *et al.* (eds.), *Change in Rural America*. St. Louis: C.V. Mosby.

_____. 1979. "Selected farm structural type characteristics: recent trends, causes, implications, and research needs." Paper presented at the Phase II Conference of the National Rural Center Small Farms Project, University of Nebraska, Lincoln.

_____. 1980. "Farm structural characteristics: recent trends, causes, implications and research needs — excerpts." Pp. 1–80 in L. Tweeten, *et al.* (eds.), *Structure of Agriculture and Information Needs Regarding Small Farms*. Washington, D.C.: National Rural Center.

Schertz, Lyle P., and others. 1979. *Another Revolution in U.S. Farming?* AER–441. Washington, D.C.: U.S. Department of Agriculture.

Schultze, Charles L. 1971. *The Distribution of Farm Subsidies*. Washington, D.C.: Brookings Institution.

Shover, John L. 1976. *First Majority—Last Minority*. De Kalb: Northern Illinois University Press.

Spitze, Robert G.F., Daryll E. Ray, Alan S. Walter, and Jerry G. West. 1980. *Public Agricultural-Food Policies and Small Farms*. Washington, D.C.: National Rural Center.

Stanton, Bernard F. 1978. "Perspective on farm size." *American Journal of Agricultural Economics* 60(December):727–737.

Tweeten, Luther, and Wallace E. Huffman. 1980. "Structural change." Pp. 1–98 in L. Tweeten, *et al.* (eds.), *Structure of Agriculture and Information Needs Regarding Small Farms*. Washington, D.C.: National Rural Center.

U.S. Department of Agriculture. 1981. *A Time to Choose*. Washington, D.C.: U.S.D.A.

U.S. Department of Commerce, Bureau of the Census. 1980. 1978 Census of Agriculture — Preliminary Report. Washington, D.C.: U.S. Government Printing Office.

Vogeler, Ingolf. 1981. *The Myth of the Family Farm*. Boulder, CO: Westview Press.

Wilkening, Eugene A. 1972. "A systems approach to policy in changing agriculture." Paper presented at the Third World Congress for Rural Sociology, Baton Rouge, Louisiana, August.

Chapter 6

European Social Theory and the Agrarian Question: Towards a Sociology of Agriculture

Howard Newby

In recent years the sociology of agriculture has assumed a much more important role than hitherto in the development of rural sociology. As I have argued elsewhere (Newby, 1982) this growing interest in the sociology of agriculture is best understood primarily as a response to the theoretical and empirical difficulties which have afflicted rural sociology, particularly in the United States, during the 1970s (see also Newby, 1980). In this context the term "sociology of agriculture" has become a rallying cry for a more critical and less empiricist approach to rural sociology. The term has come to signify something more than its literal meaning and to symbolize a "new" rural sociology which marks a radical departure from the conventional wisdom.

It is not therefore surprising that the phrase "sociology of agriculture" has come to attain a rather broad meaning. It does not, for example, simply connote a branch of occupational sociology. Indeed some writers (for example, Friedland, 1979) argue that rural sociology is *reducible to* a sociology of agriculture on the grounds that not only is the concept of "rural" a non-sociological construct, but that the conventional objects of study in rural sociology are little more than the manifest expressions of the development of productive relations in agriculture (together with other primary industries). In this sense the sociology of agriculture *is* rural sociology properly conceived. An alternative view, while in no way denying the importance of the sociology of agriculture, asserts the irreducibility of certain spatial, social and cultural aspects of rural society to the continuing development of the capitalist mode of production in agriculture. (This is my own position – see Newby, 1978, 1980, 1982.) It is not the purpose of this paper to resolve these differences, but rather to demonstrate how the discussion of these issues involves an analysis of the relationship between economy and society which was central to the development of sociology as a distinctive intellectual discipline in nineteenth-century Europe.

It is not, however, the purpose of this paper to delve into some unproductive sociological antiquarianism. On the contrary it will be argued that while recent interest in the sociology of agriculture has led rural sociologists to pay more attention to classical sociological theory and political economy, we have more to learn from the methodological example of theorists like Marx, Weber and Kautsky than from the precise content of their empirical descriptions and assumptions about the future of capitalist development in agriculture. This is because classical sociology in nineteenth-century Europe was concerned primarily with establishing a valid theory of *industrial* society (*cf*. Giddens, 1976). Agriculture was examined, if at all, only as a "background" factor – a historical backdrop from which industrial capitalism emerged. By extension "the rural" was viewed as pre-industrial, pre-capitalist and frequently as backward and *residual*. Rural sociology in the twentieth century has undoubtedly suffered from this. The comparative neglect of agricultural and rural matters by the nineteenth-century founding fathers provided an excuse for subsequent rural sociologists to ignore the contributions of the classical theorists and in particular to ignore the example they set in combining theory and method in the analysis of problems that were both socially *and* sociologically relevant.

The conscious affiliation of many rural sociologists to the sociology of agriculture has brought about some changes in this state of affairs. For example, rural sociology has to some extent shared in the growth of interest in Marxism among graduate students and junior faculty in American sociology in the recent past. In addition there has been an increasing integration between the sociology of development – where Marxism has been well-nigh hegemonic as a theoretical stance since the late 1960s – and rural sociology, as conceptualizations originally developed in the analysis of agrarian development in the less developed countries have been imported into the rural sociology of advanced industrial societies (see, for example, Friedmann, 1978, 1980). This has occurred not merely by drawing parallels between the underdeveloped world today and the processes of development in nineteenth-century industrial societies, but rather by demonstrating how the two are inextricably linked. Insofar as the developed and underdeveloped societies are but two manifestations of a single global process then it has clearly made sense to extend the analysis of the sociology of development to the former as well as the latter. Indeed sociologists of development have been driven to examining capitalist development in the agriculture of developed societies as much on empirical as theoretical grounds – as in the growing literature concerning the global dominance of transnational corporations in food, fiber and beverage production (United Nations, 1980). All of this has reasserted the relevance to rural sociologists of issues that were raised in classical political economy. The aim of this paper, then, is to investigate the uses and limitations of nineteenth-century political economy and sociological theory for contemporary concerns with a new sociology of agriculture.

Marx and Agrarian Capitalism

As indicated above, the sociology of agriculture has been extensively influenced by a growing interest in Marxism within American sociology (see, for example, many of the essays contained in Buttel and Newby, 1980; Friedland, *et al.*, 1981). However, as many writers have observed, a literal application of Marx's theory of capitalist development soon runs into difficulty when confronted by the "awkward class" (Shanin, 1972) of peasants, family farmers and other independent agricultural cultivators whose continuing existence has historically presented one of the greatest anomalies to Marx's "polarization thesis" of the class structure under capitalism, not to mention a profound political difficulty when it came to engineering a socialist transformation in the countryside. Marx's own early contempt for the peasantry is notorious. Apart from the unflattering analogy with "sacks of potatoes," Marx held little faith in their revolutionary capacity and – in a typical urban intellectual's statement of the nineteenth century – acknowledged their backward and residual historical role: they were "non-existent, historically speaking." In his earlier writings, and especially in the aftermath of 1848, Marx gave more than enough ammunition to his critics, who have cited not merely his lack of personal sympathy for the plight of the peasantry, but have used this to label Marx as a deterministic evolutionist whose theories are without relevance for the analysis of contemporary agrarian problems. Perhaps more than any other single item Mitrany's polemic (Mitrany, 1951) has been responsible for perpetuating this view.

Up until the 1960s Marxist writers therefore paid little attention to agriculture or the countryside. Then the successful peasant revolutions in China, Cuba and Vietnam, together with the widespread emergence of peasant-based liberation movements across the Third World, forced a reconsideration. Numerous attempts were made to revise Marxist theory for specifically Third World, colonial and agrarian conditions. Indeed, in the wake of the collapse of functionalism, Marxist theory became *de rigeur* in the sociology of development with debate becoming increasingly focused upon conceptual difficulties *within* a broadly Marxist framework (see, for example, the summaries presented in Ennew, *et al.*, 1976; Long, 1977; Taylor, 1979; Roxborough, 1980). Consequently in recent years the later writings of Marx, in which he was more sympathetic to the peasantry, have been re-discovered and the overall assessment revised (Duggett, 1974, and, most recently, Hussein and Tribe, 1981a, 1981b).

The problems of directly applying Marx's writings on capitalism to an analysis of capitalist agriculture are not, however, reducible to a Mitrany-like appraisal of whether Marx was for or against the peasantry. The difficulties run far deeper than this, and while they have been partially acknowledged by recent analysts (e.g. Mann and Dickinson, 1978; Goss, *et al.*, 1980; Friedland, *et al.*, 1981), they have yet to be overcome. They arise from the fact that Marx was never concerned with developing a theory of agrarian capitalism for its own sake, but rather with developing a theory of industrial capitalism, within which agriculture could be subsumed. Thus in *Capital* (especially volumes one and three) Marx (1964) writes at considerable length on the growth of capitalist agriculture in Britain, but for wholly ulterior purposes. Marx is only concerned with agrarian capitalism insofar as it accounts for the rise of industrial capitalism and insofar as it illustrates the transition from feudalism and the rise of a distinctively capitalist class structure and set of capitalist social relations. These happened, as a matter of historical fact, to occur first in British agriculture and as a matter of empirical necessity Marx is therefore forced to investigate this phenomenon. But Marx's theory of capitalist development does not rest upon this empirical analysis; nor could it, for even if Marx were to adopt such an empiricist strategy, it would lead to severe flaws in the theory of industrial capitalism which was his principal goal. As will become clear below, precepts gained from an analysis of agrarian capitalism cannot be applied to industrial capitalism nor *vice versa:* the peculiarities of the conditions of production in agriculture require a wholly distinctive analysis.

Much conceptual huffing and puffing has since been expended in trying to account for the "anomaly" of the agrarian class structure within capitalist society as a whole. It was prevalent at the turn of the century when the "agrarian question" was debated in German Social Democratic politics and in the Soviet Union after the Bolshevik Revoution. It remains in current discussion within the sociology of agriculture whereby the persistence of the "family farm," or the "small farmer" or the "peasantry" is deemed necessary to explain *away* without serious challenge to a theory of capitalism designed to account for very different, and even divergent development conditions. Indeed it is necessary to construct an analysis of capitalist agriculture which is, if not *sui generis,* then at least takes the conditions of agrarian production as a starting point rather than try to squeeze a distorted analysis into an overriding schema which is inappropriate to begin with. This suggests that classical Marxism can be utilized as a methodology and for its ability to sensitize the investigator to a set of issues rather than as a theory which can be directly applied.

The dangers of a literal appplication of Marx are further exemplified when some of the assumptions which he made concerning capitalist agriculture in Britain are considered further. Not only is Marx's analysis a kind of historical prologue to his theory of industrial capitalism, but British, and particularly English, agriculture is taken to be *prototypical.* The development of agrarian capitalism in England would, Marx assumed, eventually be followed elsewhere and the characteristic tri-partite class structure of rural England (landowners, tenant farmers, landless farm labourers) was the shape of things to come as agrarian capitalism was ushered in across Europe. With the benefit of hindsight it is possible to recognize the falsity of this assumption. The English situation, far from being prototypical, has turned out to be virtually unique. It is unique in that *only* in England was the peasantry abolished *before* the rise of industrialism. Thus today Britain and its white-settler colonies (Canada, Australia, New Zealand and — for these purposes — the United States) are distinguished by their very absence of a peasantry, whereas in virtually every other country in the world the peasantry has survived the onslaught of *subsequent* industrialization. The value of the "English model" of agrarian development is therefore limited in the extreme. It is the persistence, not the disappearance, of the peasantry which has turned out to be the most distinctive feature of agricultural capitalism.

Where Marx's analysis is clearly most useful, however, is in his analysis of rent in volume three of *Capital*. It is via his theory of rent that Marx is able to account for the existence of a landowning class conceptually distinct from the bourgeoisie and proletariat in capitalist societies. In recent years Marx's theory of rent has been the subject of extensive debate, not only within rural sociology, but within urban and regional studies (see, for example, Harvey, 1973, 1974; Edel, 1974, 1976; Ball, 1977, 1980; Fine 1979, 1980; Tribe, 1977; Massey and Catalano, 1978). There is not the space in a paper of this length to present the details of Marx's theory, nor the main lines of the contemporary debates on its validity. However, a number of general conclusions can be drawn about the relevance of recent arguments in constructing a modern sociology of agriculture.

First, as Tribe (1977) has clearly demonstrated, there lies at the heart of Marx's discussion an ambiguity concerning the conditions of existence of a landowning class within capitalist relations of production. For, on the one hand, Marx argues in *Theories of Surplus Value* that the landowning class is not a definitive aspect of capitalist society in the way in which the bourgeoisie and proletariat are, and thus he regarded the nationalization of land as a social democratic political demand which removes an important barrier to *laissez-faire* capitalism. He refers variously to the landowning class as "a useless superfetation in the industrial world," "a transitory historical necessity" and "totally superfluous" (Marx, 1967:38). On the other hand Marx's treatment of capitalist agriculture in volume three of *Capital* states clearly that the form of property in which ground rent appears is that of capitalistically transformed landed property and that capitalist relations in agriculture imply the presence of wage labor, farmer/capitalists *and* landowners who are paid contractually for permission to employ their capital in this sphere (1964:631). The first position suggests that the presence of a landowning class is a "just-so" historical fact; the second that it is intrinsic to capitalist agriculture. Since, as I have argued elsewhere (Newby,

1980) and will return to below, a consideration of landownership must be central to any sociology of agriculture, this is not an arcane theoretical problem. In many cases the structure of rural society can be derived directly from the structure of landownership (Stinchcombe, 1961); thus how the structure of landownership is organized, and its own conditions of existence and dynamic of change, are matters of considerable importance. Marx—and before him, Ricardo—pointed to the questions we should be asking without providing wholly satisfactory answers.

Secondly, it is worth emphasizing Murray's observation that: "unfortunately our general theory of the relation of landed property to capitalist development has not been developed in tandem. What is required is not merely the analysis of the capital-labor relation in [the agriculture, urban property and mineral production] sectors, but the capital-land relation. We need a general theory of modern landed property, and its concomitant, a theory of rent." (1977:100)

Why is this necessary and why, for example, is it insufficient to rely solely on a "comparative analysis of production systems" in which the social categories of organization are fundamentally the same as in manufacturing industries as Friedland, *et al.* (1981:138) suggest? It is necessary because capitalist agriculture has hitherto been unable to break the dependence of production on specific geographical locations and, given the differential fertilities and non-reproducibility of land, there is no general tendency for value to fall to the level of the least-cost producer as there is in other sectors (*cf.* Murray, 1978:28). This is not to say that capitalist agriculture does not *attempt* to break this dependence: indeed enormous research effort is expended on precisely this problem. But wherever agriculture is dependent upon land, and on specific plots of land, for successful production then ground rent is perpetuated. Where these conditions are overcome (as in much "factory" farming) then land, and therefore ground rent, loses its significance.

It follows from these two points that there lies within Marx's theory of capitalist development a theory of agrarian capitalism which is wholly distinctive. Marx did not acknowledge this. Indeed he tried to subsume agrarian capitalism under industrial capitalism. His contradictory analysis of rent results from this misguided attempt. Nevertheless once agriculture is recognized as requiring a distinctive theory of development then there is no longer any requirement for Marx's analysis of the class structure in agriculture (as in *Capital*, volume three) to conform to his analysis of the class structure in manufacturing industry. Similar comments would apply to other productive activity where land is a significant factor of production. It is on this distinctiveness that a sociology of agriculture can be constructed but the acceptance of this argument merely leads to the much greater problem which Murray indicates: we need a general theory of modern landed property.

Weber and Rationalization

Like Marx, Weber was concerned to develop a theory of *industrial* capitalism,
despite the fact that, as is well known, his model of industrial capitalism
departed from that of his predecessor in several significant respects. Weber
recognized the modern corporation, rather than the family proprietorship, as
the archetypal unit of production under capitalism: hence his concern with
bureaucracy, goal displacement, etc. Weber's theories were clearly affected by the
experience of German (rather than British) industrialization, involving a
dirigiste Prussian state and, most significantly for the purposes of this paper, the
participation of a Junker landowning class. Weber's earliest investigations dealt
with the commercialization of the Junker estates and elsewhere he offered an
"agricultural sociology" of ancient empires (Gerth and Mills, 1948:Chs. 14, 15;
Weber, 1976), but as so often in Weber's writings his treatment of agrarian
capitalism was piecemeal and diffuse, demanding much inference and *post hoc*
reconstruction. Nevertheless Weber's examination of the peculiarities of German
capitalist development does lead him to an awareness of the distinctive qualities
of continental European, as opposed to British and North American, agrarian
capitalism. Thus we find in Weber's writings an abandonment of the "English
model" of agrarian development favored by Marx and an embryonic discussion
of the fate of the peasantry which was later to dominate German social democratic politics. There is in this sense a substantive, though not a theoretical,
continuity, present in the work of Weber and the subsequent writings of
Kautsky (*cf.*, Banaji, 1980).

Drawing a comparison between Germany and the United States, Weber is
at pains to point out that it is the structure of landholding which is a determining
influence in the rural social structure:

> [T]he manner in which the land is distributed becomes of determining importance
> for the differentiation of the society and for all economic and political conditions of
> the country. Because of the close congestion of the inhabitants and the lower valua-
> tion of the raw labor force, the possibility of quickly acquiring estates which have not
> been inherited is limited. Thus social differentiation is necessarily fixed . . . (Gerth
> and Mills, 1948:364)

In the manner of the classical economists like Ricardo, Weber also recognizes the
effects of the "natural monopoly" in land and the limitations which this places
upon agriculture pursuing the same form of capitalist development as manufacturing industry:

> The importance of technical revolutions in agricultural production is diminished by
> . . . the stronger natural limits and conditions of production, and by the more
> constant limitations of the quality and quantity of the means of production. In spite

of technical progress, rural production can be revolutionized least by a purely rational division and combination of labor, by acceleration of the turnover of capital, and by substituting inorganic raw materials and mechanical means of production for organic raw materials and labor forces. The power of tradition inevitably predominates in agriculture; it creates and maintains types of rural population on the European Continent which do not exist in a new country, such as the United States; to these types belongs, first of all, the European peasant. (Gerth and Mills, 1948:364–365)

The impact of capitalism on the European peasantry is not to displace it, but to transform it:

The former peasant is thus transformed into a laborer who owns his own means of production . . . He maintains his independence because of the intensity and high quality of his work, which is increased by his private interest in it and his adaptability of it to the demands of the local market. These factors give him an economic superiority, which continues, even where agriculture on a large scale could technically predominate . . . This, again is only possible because of the great importance of the natural conditions of production in agriculture—its being bound to place, time and organic means of work. . . Wherever the conditions of a specific economic superiority of small farming do not exist, because the qualitative importance of self-responsible work is replaced by the importance of capital, there the old peasant struggles for his existence as a hireling of capital. (Gerth and Mills, 1948:367, 368)

In these passages we can observe Weber groping towards a distinctive sociology of agriculture, but, as is also clear, most of his comments are *ad hoc* and descriptive. There are few signs here of a theoretical understanding of the political economy of agricultural development. Instead Weber, characteristically, emphasizes the clash between the aggressive economism of capitalist forces and the traditionalism and inertia of the peasantry. In Weber's colorful phraseology, "The thousands of years of the past struggle against the invasion of the capitalistic spirit" (Gerth and Mills, 1948:367). What fascinates Weber is the clash of *cultures* that this involves. He is far more interested in the cultural transformation of rural society that results from capitalist penetration than he is with developing a political economy of agrarian capitalism itself. This, of course, is not surprising, since Weber's whole conception of sociology involved the denial of discernible "laws" of capitalist development. His political economy always remains implicit rather than explicit, although it is certainly feasible to suggest that he shared many of the assumptions of classical and neoclassical economics. His differences with contemporary economists lay not so much in their particular explanations but with their refusal to regard these as historically contingent (see Marshall, 1982:Ch. 2). However, Weber's opposition to historically invariant "laws" was not altogether comprehensive. This is clear from his own discussion of the concept of rationalization.

A theme which runs through the whole of Weber's work is that the process of rationalization progressively restricts the realm of independent action. Individuals are increasingly *forced* to adopt "rational" behavior, trapped in an "iron cage" where action other than that which is formally rational is no longer feasible. It is here that Weber's affinity with classical economics is most apparent, even though Weber accepts the growth of formal rationality with resigned inevitability rather than personal identification. Indeed the triumph of formal rationality is reflected in what Weber calls the "economization" of life: the rational calculation of means and ends. Weber accepts the economists' views that these ends are best measured in monetary terms. He also accepts their belief that technical efficiency can be equated with formal rationality and thus that the capitalist enterprise is technically superior to peasant and other pre-capitalist types of farm organization. As the passages cited above indicate, Weber offers a model of a dual farming economy—a technically superior and rapacious capitalist sector squeezing out of production small peasant farms whose only protection against marginalization is their ability to adapt to areas of production where there are few economies of scale and where agriculture is less capital-intensive. While Weber identifies the sources of peasant resistance to the rationalization of agriculture, however, he is in no doubt that this constitutes merely the postponement of the inevitable. The technical superiority of capitalist agriculture will ensure its ultimate victory over the forces of traditionalism in the countryside.

Insofar, then, as Weber accepts the conventional economists' account of the superior technical efficiency of large-scale agriculture and that such efficiency can be costed in terms of market prices, then he is vulnerable to equally conventional sociological critiques of classical economics—many of which, ironically, Weber would acknowledge. For example, the fact that what constitutes "rational" economic behavior is itself dependent upon a set of *antecedent* social conditions is recognized by Weber in his writings on the origins of capitalism. Similarly Weber does not recognize that peasants and small farmers might be equally "rational" in their behavior—in the sense that they are equally calculative *vis-a-vis* the market conditions that confront them—rather than a traditionalistic residue. it is possible to discern here in Weber's unflattering assumptions concerning peasant rationality the same misapprehensions which afflicted Marx. For Weber, too, the peasant is "non-existent, historically speaking." The small farmer or peasant is an *anomaly,* whose persistence needs to be explained away by reference to exceptional or "irrational" features of this culture and behavior. The crucial flaw in Weber's sociology of agriculture is thus symbolized not merely by the persistence of the small farming sector within agrarian capitalism, but in its ability to reproduce itself over several generations. The small farm sector has failed to be not only proletarianized, but also rationalized, out of existence.

Kautsky and the Agrarian Question

When, in 1899, Karl Kautsky published his important revisionist thesis of Marx, *The Agrarian Question,* it was to a background of contemporary political events which included a fierce controversy within the German Social Democratic Party (SPD) over the class position of the peasantry. This in turn was based upon the acknowledgement that the peasantry, far from disappearing, were persisting as a relatively permanent feature of rural society (at least in south and west Germany) and that a revision of Marx's assumptions was therefore overdue. Kautsky's fundamental argument was that Marx had correctly identified the *general* tendencies inherent in the capitalist mode of production, but that there are countervailing factors which prevent these tendencies being realized in particular circumstances (Hussein and Tribe, 1981a:104–106). Agriculture contains a number of features which favor the presence of these countervailing factors. *The Agrarian Question* was thus Kautsky's attempt to substantiate and elaborate the claim that agriculture possessed its own laws of capitalist development that were different from those of industry. However, it is worth pointing out, in the light of more recent discussions of the revolutionary potential of the peasantry as a class, that Kautsky regarded the distinctiveness of agrarian development as being of only limited political significance. For Kautsky industry and urban areas were the sites of major political and economic change and from where changes in agriculture initially originated (Hussein and Tribe, 1981a:107).

While Kautsky is concerned to establish the peculiarities of capitalist development in agriculture, he also notes some of the similarities with the development of capitalism in industry. There is, he argues, a steady extension of capitalist production, proletarianization and even an increasing concentration of property in means of production. But their *form* is different in agriculture. The extension of capitalism involves not so much an extension of the area occupied by capitalist farms, but vertical and horizontal integration by capitalist farmers into food processing and agribusiness. Similarly proletarianization takes a specific form in agriculture: not so much the dispossession of producers from their means of production but *the differentiation of the peasant household.* Where a peasant family found that it did not have enough land to sustain itself under existing market conditions, it sells labor rather than agricultural commodities, with the latter becoming a household activity for the purposes of supplementing the family income. In other words, the process of proletarianization is marked by the emergence of the worker-peasant, peasant-worker or part-time farmer (the modern nomenclature varies). Thus, Kautsky points out, the proletarianization of the peasant is *not* necessarily accompanied, as Marx assumed, by the disappearance of units of production organized along non-capitalist lines.

The peasant/family farmer/simple commodity producer is therefore not regarded by Kautsky as an anomaly under modern economic conditions. Furthermore Kautsky argues that the relationship between capitalist and peasant farms is not contradictory but complementary. The latter sell labor to the former during certain stages of the life cycle (*cf.* Carter, 1979), while specializing only in the production of labor-intensive commodities. This complementarity is of great significance for it implies the absence of the mechanism — market competition — whereby both Marx and Weber assumed that large-scale capitalist agriculture would become dominant. In this context proletarianization does not take a form which implies the disappearance of pre-capitalist forms of production. This opens the way for the co-existence of large-scale capitalist farms on the one hand and simple commodity producers on the other in a manner which does not threaten the existence of the latter (see also Friedmann, 1978, 1980). Whereas Marx had assumed that the process of proletarianization would accompany the destruction of pre-capitalist organizations in agriculture, Kautsky separates these two processes. This was a significant departure from what had hitherto been taken for granted in Marxist analysis, but it also represented a considerable breakthrough in the understanding of the processes at work in agrarian capital-ist development. However, as Hussein and Tribe have pointed out (1981a: 108–109), the next obvious question — what *is* the mechanism by which pre-capitalist organizations of production are destroyed in agriculture? — was never answered by Kautsky. One further point is worth noting: since the differentiated peasant household both sells labor *and* owns land its proletarianization is unlikely to have the same consequences as those which Marx predicted for the industrial proletariat. Once again the distinctive features of capitalist develop-ment in agriculture engender social effects which cannot be equated with those of industrial capitalism.

Kautsky also turns his attention to landownership and considers the observable tendencies in the ownership of property in land. In doing so he is again insistent on separating two processes: the concentration of landownership and tendencies in commodity production in agriculture. Kautsky argues that the concentration of landownership increases due to indebtedness among the peasantry and the necessity of new purchases being financed by mortgages. *De facto* control passes by this means into the hands of finance capital. But this process is not necessarily accompanied by the consolidation of small farms into large farms. For the reasons discussed above small farms may continue to survive, even though landownership may become increasingly concentrated. What regulates the concentration of *farming* is the degree of technical efficiency by which large farms may establish their superiority in a competitive market. Kautsky questioned what Marx and Weber had taken for granted: that large farms were inherently more technically efficient than small farms. As Kautsky points out, size is merely one element in the many factors which contribute to overall efficiency. Thus, for Kautsky, the capitalist character of agriculture is marked not by the scale of units of production but by the separation of the ownership of land from its cultivation — a view shared by Marx from his exami-nation of English agriculture.

Summarizing Kautsky's argument thus far, we may take note that he is concerned to:

- separate the process of proletarianization from the destruction of precapitalist forms of organization in agriculture;
- separate tendencies in landownership from those in commodity production.

The peasant/small farmer is guaranteed a modicum of survival by:

- the transformation of the peasant household;
- the dependency of large farms on small farms as a supply of labor—with small farmers thus selling labor as well as commodities;
- withdrawal from direct competition in the market for commodities;
- specialization on those commodities with least economies of scale;
- the nonextendability and immovability of agricultural land which hinders the expansion of large farms unless neighboring units can be forced or persuaded to sell.

Consequently, Kautsky argues, capitalism develops in agriculture not so much by large farms taking over small farms, but by circumventing them through the creation of "latifundia"—Kautsky's confusing term for vertical integration into food processing and similar ancillary agribusiness activities. Capitalist agriculture consists not of integrated units of production, but of integrated units of ownership and management composed of a number of units of production. Kautsky foresees the growth of vertically integrated, food production conglomerates within which agriculture—i.e. cultivation—forms only one part of the process of food manufacture. It is in this way, Kautsky argues, that agriculture and industry will eventually reconverge in their developmental logic.

Conclusions

What lessons can be drawn from this brief excursion into nineteenth-century European social theory? The first, and most general point to make is that the theories of Marx, Weber and Kautsky were developed in a particular historical context and as part of on-going political debates which shaped their presentation and their "value-orientation." Their theories are not entirely polemical, but neither are they abstract or timeless. The danger is that in the enthusiasm to inject some much-needed theoretical holism into rural sociology, North American scholars will attempt a too-literal application of these theories to contemporary conditions. These writers deserve attention for the example they set, for their method and for their insights. They are less exemplary as predictors of empirical reality.

With the benefit of hindsight their weaknesses have become apparent. Their limited and partial appraisal of agrarian development has already been emphasized. The center of attention for each of these writers was urban, industrial capitalism and not the countryside. Marx places too much faith in the English model; Weber in the technical superiority of large-scale production. Kautsky offers much the most sophisticated analysis and draws a number of crucial distinctions, but many of his empirical predictions have gone awry. The creation of a sociology of agriculture is thus unlikely to be achieved by a simple rediscovery of overlooked or misinterpreted classical texts.

Nevertheless these writers do point to the kind of *questions* which a "new" sociology of agriculture should be concerned with, even if they do not adequately furnish the answers. At the very least they suggest an extensive and fruitful research agenda. They also, sometimes inadvertently, establish the validity of a separate and distinctive sociology of agriculture because of the importance of *land* as a factor of production. It need hardly be stressed that we have no adequate theorization of landownership in the way in which theories of capital are available. Are we to accept the pattern of landholding as a *sui generis* historical fact which is not amenable to sociological theorizing? If so, the implications for a sociology of agriculture are profound indeed, for it would suggest that no comparative sociology of land tenure and land use would be possible. The answer must surely be, however, that rural sociologists must direct far more attention to landholding in order to establish a distinctive sociology of agriculture. On this, at least, Marx, Weber and Kautsky point the way.

Clearly, however, the construction of a sociology of landownership will not, in itself, be sufficient. The relationship between landownership and commodity production must also be analyzed for it is the articulation *between* these two—rather than the characteristics of merely one or the other—which is decisive in understanding the development of modern agriculture. This suggests the need for a comparative political economy of agriculture, but one which takes account of this articulation. Unfortunately, however, this is an issue not addressed, for entirely valid reasons, by analysts of the political economy of industrial capitalist development. It is also important to retain a degree of flexibility in such an analysis. To be specific: it may be appropriate to treat certain branches of agriculture (intensive livestock and horticulture production, for example) *as if* it were industrial production, whereas in other branches (dairying most notably) such an analogy is inappropriate. Therefore if analyses drawn from the political economy of industrial production are not automatically transferable to agriculture, neither are they completely irrelevant. But they will be *modified*—more or less according to the type of commodity production—by the necessity of accounting for landownership.

These proposals may seem unduly economistic. It is not, however, the intention to turn the sociology of agriculture into a branch of agricultural economics. The emphasis given in this paper to economic factors is, rather, an attempt to redress the balance from the almost total disinterest in such matters which characterized rural sociology for so long. It is, therefore, important to conclude with a re-assertion of the primacy of the social. At a very abstract level it might be taken as axiomatic by sociologists (certainly by Marx, Weber and Kautsky) that economic relations of a particular kind require a set of antecedent social conditions in which to flourish. But rural sociology does not need to justify itself only on such abstract grounds. Within a capitalist society land *is* capital. Land may not be mobile, but capital is. Why then is capital-in-land not transferred with the same alacrity in response to market conditions which characterizes finance or industrial capital? The answer to this question is frequently deeply social, relating to the values, customs, patterns of kinship and household structures which impinge upon the ownership of land. Here, at least, rural sociology is not reducible to agricultural economics. Indeed a considerable amount of information on these issues already exists, but needs to be codified and somewhat reinterpreted around the issues raised in this paper. Not only must the sociology of agriculture contain a comparative political economy, but this political economy must contain a sociology of risk and entrepreneurship.

References

Ball, M.J. 1977. "Differential rent and the role of landed property." *International Journal of Urban and Regional Research* 1(October):380–403.

_____. 1980. "On Marx's theory of agricultural rent: a reply to Fine." *Economy and Society* 9(August):304–326.

Banaji, Jarius. 1980. "A summary of selected parts of Kautsky's *The Agrarian Question*." Pp. 39–82 in Buttel and Newby (eds.), *The Rural Sociology of the Advanced Societies: Critical Perspectives.* Montclair, NJ: Allenheld Osman.

Buttel, Frederick H. and Howard Newby (eds.). 1980. *The Rural Sociology of the Advanced Societies: Critical Perspectives.* Montclair, NJ: Allenheld Osman.

Carter, Ian. 1979. *Farm Life in Northeast Scotland, 1840–1914.* Edinburgh: John Donald.

Duggett, Michael. 1974. "Marx on peasants." *Journal of Peasant Studies* 2(April): 159–182.

Edel, M. 1974. "The theory of rent in radical economics." Working paper 12, Boston Studies in Urban Political Economy.

_____. 1976. "Marx's theory of rent: urban applications." Pp. 7–23 in CSE (ed.), *Housing and Class in Britain.* London: CSE.

Ennew, Judith; Paul Q. Hirst; and Keith Tribe. 1976. "'Peasantry' as an economic category." *Journal of Peasant Studies* 4(July):295–322.

Fine, Ben. 1979. "On Marx's theory of agricultural rent." *Economy and Society* 8(August):241–278.

_____. 1980. "On Marx's theory of agricultural rent: a rejoinder." *Economy and Society* 9(August):327–331.

Friedland, William F. 1979. "Toward a sociology of agriculture." University of California–Santa Cruz, Department of Sociology, mimeograph.

Friedland, William F.; Amy Barton; and Robert Thomas. 1981. *Manufacturing Green Gold.* Cambridge: Cambridge University Press.

Friedmann, Harriet. 1978. "World market, state, and family farm: social bases of household production in the era of wage labor." *Comparative Studies in Society and History* 20(October):545–586.

———. 1980. "Household production and the national economy: concepts for the analysis of agrarian formations." *Journal of Peasant Studies* 7(January):158–184.

Gerth, H.H., and C.W. Mills. 1948. *From Max Weber.* London: Routledge and Kegan Paul.

Giddens, Anthony. 1976. "Classical social theory and modern sociology." *American Journal of Sociology* 81(January):703–729.

Goss, Kevin F.; Richard D. Rodefeld; and Frederick H. Buttel. 1980. "Capitalist agricultural development and the exploitation of the propertied laborer." Pp. 83–132 in Buttel and Newby (eds.), *The Rural Sociology of the Advanced Societies: Critical Perspectives.* Montclair, NJ: Allenheld Osman.

Harvey, David. 1973. *Social Justice and the City.* London: Edward Arnold.

———. 1974. "Class-monopoly rent, finance capital and the urban revolution." *Regional Studies* 8(November):239–255.

Hussein, Athur, and Keith Tribe. 1981a. *Marxism and the Agrarian Question, Vol. 1: German Social Democracy and the Peasantry, 1890–1907.* London: Macmillan.

———. 1981b. *Marxism and the Agrarian Question, Vol. 2: Russian Marxism and the Peasantry, 1861–1930.* London: Macmillan.

Long, Norman. 1977. *An Introduction to the Sociology of Rural Development.* London: Tavistock.

Mann, Susan A., and James H. Dickinson. 1978. "Obstacles to the development of a capitalist agriculture." *Journal of Peasant Studies* 5(July):466–481.

Marshall, Gordon. 1982. *In Search of the Spirit of Capitalism.* London: Hutchinson.

Massey, D., and A. Catalano. 1978. *Capital and Land.* London: Edward Arnold.

Marx, Karl. 1964. *Capital,* Vol. III. London: Lawrence and Wishart.

———. 1967. *Theories of Surplus Value.* London: Lawrence and Wishart.

Mitrany, D. 1951. *Marx Against the Peasant.* London: Weidenfeld and Micolson.

Murray, Robin. 1977. "Value and theory of rent: part one." *Capital and Class* 8:100–122.

———. 1978. "Value and theory of rent: part two." *Capital and Class* 9:11–33.

Newby, Howard. 1978. "The rural sociology of advanced capitalist societies." Pp. 3–30 in H. Newby (ed.), *International Perspectives in Rural Sociology.* London: Wiley.

———. 1980. "Rural sociology—a trend report." *Current Sociology* 28 (Spring): 1–141.

———. 1982. "Rural sociology in these times." *The American Sociologist* 17(May): 60–70.

Roxborough, Ian. 1980. *The Sociology of Development.* London: Macmillan.

Shanin, Teodor. 1972. *The Awkward Class.* London: Oxford University Press.

Stinchcombe, A. 1961. "Agricultural enterprise and rural class relations." *American Journal of Sociology* 67(September):165–176.

Taylor, John. 1979. *From Modernisation to Modes of Production.* London: Macmillan.

Tribe, Keith. 1977. "Economic property and the theorisation of ground rent." *Economy and Society* 6(February):69–88.

United Nations. 1980. *Transnational Corporations in Food and Beverage Processing.* New York: U.N. Center on Transnational Corporations.

Weber, Max. 1976. *The Agrarian Sociology of Ancient Civilizations.* London: New Left Books.

Chapter 7

Policy Evolution and Current Crisis in Polish Agriculture

Boguslaw Galeski

Introduction

The main idea in the doctrine legitimizing the claims of communist parties to political power, particularly in developing countries, is unity between the relations of production and productive forces; forms of ownership of the means of production must conform to technology. In simplified versions of Marxism, private, capitalist ownership of the means of production is seen as the main source of economic crisis in the capitalist system (*cf.* Stalin, 1940). It stands as the primary obstacle to the development of technology and consequently to economic growth, especially in countries on the periphery of world capitalism.

The establishment of a communist state is presented to the people in developing countries as the best strategy for accelerating economic growth. Complete state control of all national resources, it is argued, will enable them to "jump" the evolutionary stages of capitalist development and achieve a more prompt eradication of poverty and a more secure position in the international arena.

Of course, belief in the superior efficiency of the autocratic state is not limited to Marxists. The idea that a state controlled national economy can mobilize all available resources more easily and quickly than western democracies is expressed quite often by non-Marxists. It is used to justify military dictatorships

The author expresses appreciation to Gene F. Summers for his extensive editorial assistance. This manuscript was written while the author was visiting the University of Manchester, and therefore documentation is somewhat limited due to unavailability of materials.

in developing countries and the violations of human rights as the price that must be paid for technological modernization and economic growth.

Poland during the inter-war years could hardly be called a developing country, in the current sense of the term. However, after regaining national independence in 1918, Poland urgently needed industrial investments to integrate her economy and to exploit her resources in a more balanced fashion. Although the Great Depression did not offer much chance for that, by the mid-thirties the start of economic recovery was evident. To continue the process, land reform was urgently needed to reduce rural poverty and underemployment. It was needed also as a pre-condition for the diffusion of modern agricultural technology. But legislation to implement the process was delayed by the opposition of a politically influential landed aristocracy. And the process was interrupted by the war.

At the end of World War II, Poland became a communist state, part of the Soviet bloc. The monopoly of the autocratic state, in control of all basic national resources, was firmly established. Accepting doctrinal principles and following the Soviet model of development, the communist government introduced an ambitious program of industrialization and devised a policy intended to achieve organizational and technological "modernization" of agriculture.

The majority of farmers accepted land reform but were strongly opposed to enforced collectivization and thus Polish agriculture remained predominantly private. However, that did not restrict the state's control over other factors of agricultural production. The state retained its monopoly in the importation, production and distribution of agricultural technology and its monopoly over the market, strictly controlling the level of farmers' income and their purchasing power. State investments were concentrated in the state-owned farms which were intended to demonstrate the superiority of the socialist forms of production, technologically and economically.

Thus, the basic conditions for greater productivity in agriculture, and the whole economy were fulfilled, according to the doctrinal formula. But the expected productivity did not materialize. The rate of growth of the total economy was slow, importing of agricultural products continued to rise, and by the end of the 1970s, the decline of both industry and agriculture was evident.

The failure of the strategy of economic development and particularly the failure of agricultural policy can be observed not only in Poland but also in other countries of the Soviet bloc and in the Soviet Union itself. But the mechanisms of a growing crisis are more easily accessible for direct observation in Poland than elsewhere. Partly this is due to the fact that in Poland more research has been done and, therefore, more information is available. Even so, the most important information, such as the terms of trade with the Soviet Union, have never been released. However, during the political crisis in 1980–1981, the veil of optimistic propaganda was removed; basic facts came into the open which could not be denied. An analysis of the political and economic crisis is therefore easier to carry out in Poland than in any other country of the Soviet bloc.

In agriculture, Poland offers the unique opportunity to compare the performance of socialized and private forms of agricultural production. Both operated within the same socio-economic system which provided advantages for the former and disadvantages for the latter. Undoubtedly, some results should be attributed to the functioning of the Soviet bloc system rather than to the legal ownership of means of agricultural production or to the size of farms in Poland. It will be difficult to single out those factors which should be attributed to the socio-economic system in the Soviet bloc and those such as private agriculture which are peculiar to Poland. However, the Polish crisis provides an opportunity for such analysis and it is worth doing, at least, in relation to agriculture.

The Agrarian Question

The programs to implement agricultural policies adopted by communist parties and communist governments are based on the set of basic assumptions of Marxism-Leninism under the title of "the agrarian question" (Konstantinov, 1965).

There are three strategic aims of agricultural policy which are derived from basic doctrinal assumptions, translated into operational terms and implemented by communist states. One of these policy goals is to overcome the presumed inappropriateness of family farms for extended reproduction of capital, i.e., for economic growth (Kautsky, as translated by Banaji, 1976), and for the adoption of modern technology. The transformation of family farming into large-scale farming organized along industrial lines is, therefore, required as a strategic aim of the communist state.

A second policy goal is to eradicate the socio-economic basis for exploitation of agricultural families by rural and urban capitalists, and further eliminate the grounds for social differences (considered as social contradictions) between the city and the countryside. The presumed basis of such exploitation is private, capitalist ownership of the means of production. Realization of this goal, therefore, is sought through the socialization (in fact, nationalization) of all means of production including those of agricultural production.

The third policy goal is to harness the harsh uncontrolled market forces which cause economic crises, waste of resources and lead to the emergence of capitalistic relations of production (Lenin, 1921). This objective is supposed to be reached by replacing the market economy with a centrally planned economy.

Ultimate Model of Agriculture

Doctrinal assumptions and strategic aims indicated in agricultural programs imply the ultimate model of agriculture. This model is usually not described fully in information presented to the public; there is no need to antagonize farmers whose participation, or at least compliance, is required for the fulfillment of planned tasks. In addition to technicalities, public descriptions of programs usually contain only a vague reference to "socialist agriculture" as the final aim of all planned efforts. However, the model can easily be deduced from programs, textbooks of Marxism-Leninism, or specialized books on the "agrarian question" released for publication by the communist party. For Marx and Engels, for Lenin and Stalin, and for recent leaders of communist parties in the Soviet bloc, there is no doubt; the large-scale organization of production in agriculture, as in industry, is an essential condition of socialism, and of economic and technological development. The ultimate goal of the socialist transformation of agriculture is large-scale, state-owned farms, fully mechanized, employing specialized labor, and directed by central management as an integral part of the national economy. Contrary to wide-spread belief, the collective forms of ownership of the means of production in agriculture (Soviet Kolkhozes, East-European cooperatives for agricultural production, etc.), were not considered the final form of socialist agriculture, but a compromise or a transitional form only. Both in the doctrine (Stalin, 1952) and in agricultural policy promoting the concentration of agricultural production, state-owned and centrally directed units of agricultural production are characterized as "higher" stages of socialism in agriculture. True collective farms (and true cooperatives in general) which secure the control of direct producers over means, processes and results of production, were never allowed to exist in the Soviet Union and were never accepted by the doctrine of Soviet Marxism. However, the leaders of the communist parties in some countries have been allowed to design their own route to socialist agriculture. In these instances, the ultimate achievement of the final model was temporarily postponed, as in Poland, or accelerated, as in Cuba, because of special conditions. But always, the same ultimate model must be kept as the aim of agricultural policy; no deviations from it are tolerated. This model was and is accepted also by the leaders of the Polish communist party, independently of changes in political leadership and of different formulations of agricultural programs in successive periods.

Luxemburgism

The views of the pre-war Communist Party of Poland (KPP) on the agrarian question ought to be mentioned here, not because the Party or its views had any influence on the agricultural policy at that time, but because the heritage of KPP can be felt later in the program of the Communist Party and in its execution after the war. KPP was considered to be a rather marginal political group in pre-war Poland. The weakness of the party was due, according to its marxist critics, to the errors on (1) the national question, opposition and later indifference to the national independence of Poland, and (2) the peasant question, their view of the peasantry as a reactionary class and consequently their distrust of peasant radical movements. These errors were branded as "luxemburgism" (Rosa Luxemburg was one of the leaders of Social Democratic Party of the Kingdoms of Poland and Lituania [SDKPIL], predecessor of KPP) and were subject to continuous discussion in KPP during its existence. In 1938, just before the Molotov-Ribbentrop pact, KPP was dissolved by the Comintern; most of its leaders were summoned to the Soviet Union and executed there. Some survived the purge, mostly those who were imprisoned in Poland at that time, and later gained top positions in the Communist Party after the war (PPR – Policy Workers Party and later PZPR – Polish United Workers Party).

The heritage of Luxemburgism contributed to the inconsistency of agricultural policy in Poland after the war. Even though the Communist Party followed the Soviet pattern strictly in 1948-1956, it has faltered over the years. It could be said that Luxemburgism imprinted anti-peasant feelings on some Party leaders and fears of repetition of Luxemburgist errors on others; for instance, Gomulka opposed collectivization. In this way it contributed to the ambivalence of agricultural policy in Poland.

Stages of Agricultural Policy

The evolution of the agricultural program in the post-war period went through four stages. Two of these, 1944-1948 and 1948-1956, are quite similar to those in other East-European countries. After 1956 the evolution of general policy, including agricultural policy among Soviet bloc countries, was much more differentiated (Hungarian economic model, Rumanian relative independence in foreign policy, etc.). In Poland the period 1956-1970 (under the leadership of Gomulka) reflected an agricultural policy which presumed stagnation of the economy and shall be distinguished from the period of rapid technological and economic change during 1970-1980 (under the leadership of Gierek). The political and economic collapse in 1980 opened a new stage which, even after the introduction of martial law in 1981, has not yet revealed its more characteristic features, particularly in economic strategy.

1944-1948

The first stage of the Communist Party rule in Poland was devoted to gaining full political power and introducing substantial economic reforms. In industry, banking and transportation property was nationalized, in agriculture land reform was initiated. In Poland, as in other East European countries, large landholdings were expropriated by the State and 70 percent of the land distributed to landless families or the owners of farms of inadequate size, below 2 hectares (1 hectare = 2.471 acres). In eastern and central Poland large landholdings were defined as 50 hectares and in western Poland, more than 100 hectares. Thirty percent of the expropriated land was kept by the state for the creation of state farms. These large-scale farms were managed by specialists appointed by the government and their production integrated into centrally designed national economic plans.

Former agricultural laborers were the largest group to gain access to the expropriated land, 47 percent of that which was distributed, or approximately one-third of the land taken from large landowners. Most of these new land owners had no previous experience in independent farming and, therefore, met extreme difficulties in the beginning. Many were helped by the UNRRA program with acquisition of seeds and draft horses and were successful in establishing their farms. But the majority were unsuccessful as independent farmers and either voluntarily joined with others to organize collective farms or returned their land to the State and asked to be accepted as employees. During this period, the strategic aims of the Party were not indicated openly in its agricultural program, only that the final aim of the agricultural transformation was "socialism." The exact meaning of the term, the timetable and the methods of accomplishment were left undisclosed.

Some temporary collective farms were created in the former German territories of western and northern Poland where large estates sometimes were given to groups of new settlers for distribution to individual families. Initially these settlers had to share agricultural buildings (barns, cowsheds, etc.) and equipment, but the idea of joint farming apparently was acceptable to them since they often came, as a well integrated group, from the same villages in Ukrania or Bialorussia. They were given the legal status of "settler's cooperative for land distribution" on the condition that in the near future, usually five years, they would dissolve the cooperative and begin independent farming.

This apparent departure from the ultimate model of agriculture can be understood by recognizing that the Communist Party in Poland (Polish Workers' Party) was in a politically sensitive period; it had no public support. In spite of the land reform the peasantry was opposed to communism and organized several opposition political parties; the most influential was the Polish Peasant Party (PSL). The political situation was quite volatile at that time; in some parts of Poland former national guerillas were still fighting against the new government. Therefore, the Communist Party wanted to avoid encouraging the suspicions that it intended to introduce the Soviet system in agriculture. In consequence, collective farms, even those spontaneously organized by former agricultural laborers, were not permitted by the government and those that did exist (no more than 100) were declared illegal. The leaders of the Communist Party in Poland several times made official announcements that the Polish political system was not to be a "dictatorship of the proletariat" and that the Polish way to socialism, particularly in agriculture, was to be quite different from that in the Soviet Union.

During this period, Party agricultural policy was focused mainly on the distribution of land from former estates, the provision of financial help for new settlers and the management of state farms. Private farmers owned more than 85 percent of the agricultural land at that time, and those who were not involved in land reform or re-settlement were not disturbed in their work. As a result, the productive potential of Polish agriculture, damaged heavily during the war, was reconstructed in a very short time. The level of technology in private farming was quite low; the use of tractors was rare and the use of chemical fertilizers very sparse (ca. 10kg of NPK per hectare on average). Even so, by 1948 the average yield per hectare had reached the pre-war level. Projections of a relatively fast economic growth and social development appeared to be quite realistic. However, all these high expectations disappeared in the next period.

1948–1956

In this period the communist parties in the East European countries gained full political control. The former leaders of the party were denounced as "opportunistic nationalists" or "Titoists" and, with the exception of Gomulka in Poland, sentenced to death. The "dictatorship of the proletariat" was declared and new socio-economic programs of development were issued, strictly following the path taken in the late twenties by the Soviet Communist Party under the leadership of Stalin. In industry this implied huge investments in new plants, particularly in metalurgy. In agriculture the program of collectivization was announced, accompanied by political campaigns either against the rich farmers (kulaks) or against all farmers refusing to join collective farms. In Poland the program of collectivization started in 1949, much earlier than in other East European countries with the exception of Bulgaria. It was implemented using detrimental economic measures such as discriminatory taxation and requisitioning of grain and livestock. Also, harsh and repressive administrative measures were used, particularly in 1951 and 1953, including imprisonment of farmers who voiced their dissatisfaction.

Nonetheless, the collectivization in Poland proceeded rather slowly. By 1056 there still were fewer than ten thousand collective farms, incorporating no more than 10 percent of the agricultural land. Together with the campaign against "kulaks", the disruption associated with collectivization hampered the growth of agricultural production. At the same time there was a rapid growth of urban population (3 million between 1950 and 1956) creating a growth in the demand for agricultural products. In 1956, for the first time after the war, Poles experienced shortages of food on the market. This, and a shaky national economy, aggravated an already tense situation.

The political crisis of 1956 and the first massive protests of Polish workers against the communist state cannot be explained by economic factors only. The roots of the political upheaval have to be sought also in nationalistic feelings of the Poles (particularly of Polish workers), embittered by the obvious dependency on and enforced imitation of the Soviet system and of the Soviet way of life. Another important factor was the anti-religious campaign of that time, carried on in a country where more than 95 percent of the population was Catholic and where for centuries Catholicism had been an inseparable part of the system of national values.

There is a frequently voiced opinion that collectivization failed in Poland because of the strong resistance of the peasantry, but there is no evidence to support such an interpretation. It is true that the program of socialist transformation of agriculture was even less popular in Poland than in other East European countries. In no country did the peasants enthusiastically support collectivization. In Poland it was rejected not only by peasants; it was rejected by non-agricultural workers and by a large number of middle rank party officials who were mostly of peasant heritage. Moreover, the Polish peasantry was much more numerous and influential as a social force than farmers in more industrialized Czechoslovakia and East Germany. Also they were more attached to independent family farming than impoverished cottagers and tenants in less industrialized Rumania and Bulgaria.

But a much stronger resistance by the Russian peasantry was stamped out by the communist state. The fate of the Polish peasantry would have been the same had it not been for the general political crisis which followed the death of Stalin, Krushchev's speech to the Twentieth Congress, and the uprising of workers in Poznan followed by the national uprising in Hungary. These events had shaken the system of political control. Additionally, there is the very important fact that in these circumstances, the only alternative political figure able to gain public confidence was the former party leader Gomulka. As a member of KPP, Gomulka belonged to those who were deeply frustrated by "Luxemburgist" errors in dealing with the national and peasant questions who were stubbornly searching for the "specific Polish way to socialism."

Coming back to power, Gomulka ordered dissolution of most of the collective farms and declared the need for a new agricultural program. In this way Poland became an exception among East European countries, where collectivization of agriculture was later completed.

The collectivization effort had failed in Poland but other elements of the agricultural program of 1948–1956 remained basically unchanged and two were continuously strengthened in the years that followed: enlargement of the socialist sector, and tightening of the state market monopoly.

Extension of the socialist sector in agriculture was given priority in agricultural policy. In 1950 the State and collective farms represented jointly 10.3 percent of the cultivated land. By 1956 the percentage was 23.0 but fell sharply after dissolution of collective farms to 13.3 percent in 1957. After 1957 it grew steadily and by 1980 reached 24.2 percent.

The new technology which was divided between state farms and collective farms in the socialist sector from 1948 to 1956 was concentrated exclusively in state farms after 1956. Between 1950 and 1957 the number of tractors in agriculture doubled (28,000 to 56,000), but *new* tractors were sold only to state farms or to state owned tractor stations. The task of the tractor stations was to organize and to serve collective farms. After dissolution of collective farms the tractor stations were transformed into a branch of the state farming system. The machinery was taken to the state farms and the stations then served only as repair workshops. Also, chemical fertilizers and other agrochemicals were sold primarily to state and collective farms from 1948 to 1956. The use of chemical fertilizers rose from 17.7kg NPK per hectare in 1950 to 30.7kg in 1957. But these figures are computed as an average for the total cultivated area, and it can be estimated that in state and collective farms the use was about 70kg NPK per hectare since private farms remained on the previous level. The priorities given to the socialist sector were not limited to material factors of production only. Credit, services, low taxation, free medical care and pensions, etc. also were provided to those employed in the state farming systems, but not to private farmers. Programs such as research, agronomic services, rural education, and dissemination of agricultural knowledge, were designed to serve exclusively the socialist sector. Private farmers and their families became second class citizens in all those aspects of life subjected to the omnipotent state. Inspite of these priorities, however, the socialist sector failed to reach a high level of productivity. For the whole time, right to the present, its share in total agricultural production was significantly below its share of the total cultivated areas.

The second element of the agricultural program of 1948–56, which remained unchanged in the future, was total monopolization of the market for agricultural products by state agencies. For some products such as grain and meat, the system of obligatory deliveries, introduced in 1952, was continued. Other products not included in this system were covered by contract production. Beginning in 1947 agricultural cooperatives (credit unions, dairy cooperatives, fruit and vegetables cooperatives, universal cooperatives, etc.), like all cooperatives, were placed under the control of state agencies and forcibly transformed into parts of the centrally directed systems. Indeed, they ceased to be cooperatives at all, if by "cooperative" we mean a voluntary, self-governed association.

As control over economic affairs of the cooperative by its members became virtually impossible and economic benefits of membership were lost, active participation became insignificant. Management did not need more "members" and members lost all interest in the cooperative's activities. This alienation of cooperative members reached such dimensions that farmers asked in the mid-1950s whether they were members of cooperatives could not even recall. In this way, even without collectivization, the state gained full control over the pricing of commodities in private agriculture and, most importantly, over the income of farming families.

The creation of this system cannot be justified merely as promoting collectivization, or by political reasons. The basic economic reason was the ambitious program of industrialization which needed both the transfer of manpower from the countryside to industrial centers and sufficient supplies of cheap basic food for the growing number of urban dwellers. By being forced to sell their products to the state at low prices, peasants were made to pay a large part of the cost of industrialization. The market monoply, collectivization, and the threat of it, hampered private farms and contributed to the substantial breakdown of the economic motivation of agricultural producers which has never been regained. Farming families learned by their experience that hard work would not secure their economic and social position, or that of their children. The ethos of the thrifty farmer, formed mostly in the second half of 19th century, was destroyed.

In actuality the demise of productive private farms began during the period of 1948-1956 when the value of agricultural production on private farms was stagnant but still higher than on state and collective farms. The ethos of farming families conveys the belief that they cannot afford to neglect the land; they need food, they need cash, however unfavorable the prices, and they feel obliged to preserve the fertility of the land. But the usual investments into agricultural production were reduced sharply because of the unavailability to private farmers of construction materials and other inputs to the maintenance of agricultural production. Also, under the threat of collectivization there was no rational basis for making such investments and the money usually spent for that purpose was used for other family needs.

However, some farmers succeeded in adapting to the conditions created by state agricultural policy. By having personal connections with people in high positions in state administration and state owned businesses, or by simply bribing them, some farmers got access to more profitable contracts for agricultural production and to restricted materials (construction materials, fertilizers, commercial feed, etc.). Even so, they were not investing much in long-term production maintenance but rather operating to maximize profit in the short run. Thus, the breakdown of producers' motivation and the channeling of entrepreneurial talents of some agricultural producers into socially and economically detrimental activities may be regarded as unplanned social consequences of the agricultural policy of the period (1948-1956). The crisis in the morale of rural society (and society in general) came to evidence much later, particularly after 1970 but its foundations were established already in the 1948-1956 agricultural policy of the party. In this way the germs of the collapse of Polish agriculture in the late 1970s were already laid in the period of 1948-56.

The other unexpected result of the economic policy of 1948-1956 was the formation of a new, large, and constantly growing social stratum of so-called "peasant-workers" (one and one-half million families by 1982). The majority of farmers owning mini-farms (under 2 hectares) and a large number of the owners of small farms (between 2 and 5 hectares) found supplemental employment in unskilled and low-paid jobs in the growing industrial sector, mostly as construction workers. Consequently, farming became a secondary source of income, at least of cash income. The main role of the farm was to provide the family with basic food with little or nothing produced for the market.

Peasant-workers were from the very beginning regarded with dismay by political leaders, technocrats, managers of industry, and even by some agricultural economists. The peasant-workers were accused of living better than the average industrial or farm family, of buying all the attractive goods on the market, and of not participating in the social activities intended to strengthen party and national unity. Paradoxically, they were also accused by some sociologists of being the main driving force in workers strikes and riots, particularly the shipyard workers' uprising in 1970.

All these accusations have been shown to be false. The farms of peasant-workers are at least at the national average of productivity, a large number are kept highly productive. It is a matter of fact that they sell little of their produce; but they sell no less than other full-time operators of farms in the same size category (up to 5 hectares) excepting highly specialized farms. During 1948–1956 highly specialized farms of small size were very uncommon, both the incentives and the means needed for specializaton were lacking.

It appears there were at least two reasons for the widespread perception that peasant-workers were enjoying "the good life" unfairly. First, having a food supply from their farms the peasant-workers avoided the food-queue and it was for that reason they were perceived by some urban dwellers as having a "good life". The other reason was that unlike other full-time farmers, peasant-workers had access to free medical services and welfare benefits because of their lower standard of living. Even today the average standard of living of peasant-worker families is about 18–20 percent lower than that of industrial workers in general, and about 10 percent lower than middle-level farmers. This is due primarily to the fact that peasant-workers are usually unskilled and can get only the low-paying jobs. In contrast with urban industrial workers they have lower rates of job turnover and are more compliant with rules set by management. Only during harvest time do they take days of leave (with or without permission), which is much less absenteeism in total than their urban co-workers. It is also true that peasant-workers have no time for propaganda meetings or for ideological schooling. Indeed, they are not interested in the "socialist transformation of agriculture" and in this sense they do create an obstacle for the fulfillment of the agricultural program of the party.

Even though peasant-workers are an anomaly to party doctrine and disliked by many, authorities have been unable to solve the problem. Faced with the situation of a drastic shortage of housing in the cities, permanent shortage of food on the market, and lack of material incentives to raise the efficiency of labor in general, the peasant-workers cannot be dismissed. Indeed, peasant-workers have contributed significantly to the industrialization of Poland. Accepting low wages and commuting long hours, peasant-workers built most of the new industrial plants in Poland and later provided the permanent labor force.

1956-1970

The political upheaval of 1956 and the return of Gomulka to party leadership brought optimistic hopes for substantial changes. There were surely some signs of a change for the better, not only in agriculture. The enforced collectivization was stopped, the party bureaucracy on all levels was modified and officals were generally more reluctant to exert pressure on farmers. The old farmers' organizations, so called "agricultural circles", were restored and granted new and important economic functions, particularly in buying and keeping tractors and heavy farm equipment for joint use in the community. Certain amounts of highly desired industrial goods (mostly construction materials) were placed at the disposal of farmers. The system of obligatory deliveries was "temporarily" maintained, but at least some of its financial benefits (the difference between prices in the system of obligatory deliveries and prices paid by state agencies in "free" buying) were paid to the newly established Fund for Agricultural Development which provided agricultural circles with money to buy tractors. The "jump" in agricultural productivity in 1957 was accompanied by a general improvement in the economy (the admittance of private business in handicraft and services) and a better political atmosphere (agreement between State and Church, cultural "thaw", etc.). In this atmosphere, the postulate of unification of socialism with rising productivity might seem quite realistic, for a while.

The party never gave up the idea of socialist transformation of agriculture, but after the failure of collectivization, a new approach was taken. It was manifested in the postulate that the transformation should be achieved through the process of increasing productivity of private farming (Joint declaration of PZPR and ZSL—Polish United Workers Party and United Peasants Party—1957). Some social scientists regarded it from the beginning as contradictory in its content or treated it as a pure propaganda statement. Others, particularly in the period of 1957-1960, when serious debate was still tolerated, thought its achievement was possible and offered elaborate proposals for the formulation of an agricultural program for its implementation.

One of these proposals for designing a socialist version of the vertical integration of agriculture was particularly promising (*cf.* Tepicht, 1973). It was partially implemented in agricultural policy with the extension of contract production but significant results in the socialization of private farming were not achieved. In its official documents the party has never translated the postulate into a set of clear directives and, therefore, it is difficult to say how realistic it was; it has not been tested in practice.

Looking back on the meanderings of agricultural policy after 1956 it seems that the postulated union of socialist transformation and rising productivity through private farming was (or in its execution was), in fact, contradictory. Sometimes the authorities pushed socialization of agriculture making the situation of private farming unbearable; sometimes (less often) under the pressure of political and economic difficulties, they made a few concessions (mostly adjusting prices for agricultural products), and farmers breathed more freely for awhile. From today's vantage point, however, it is clear that the evolution of Polish agriculture since 1956 has not been guided by a consistent agricultural program. Lack of consistency is the only consistent feature of agricultural policy since 1956. Other factors, such as the general evolution of the economic and social situation in Poland, have been much more decisive for the performance of agriculture than agricultural policy in itself.

In 1956–1957 the program of economic reforms, similar to those introduced later in Hungary, was prepared by the specially appointed Economic Council. It called for decentralization of the rigid system of planning and management and demanded that newly formed worker-councils in industry should have a part in economic decision making. But substantial reform of the economic system was not accepted by the party; the Economic Council was dissolved shortly after 1957 and the elected members of worker-councils were replaced by nominated party functionaries. As the party regained full political control, the concessions granted during the time of political upheaval were recalled, or rather the promises of concessions were not fulfilled. The reborn production motivation of farmers was also killed. Agricultural circles were placed under the control of state agencies, the agricultural prices were kept on a low level (compared to industrial prices), the supply of agricultural technology (fertilizers, construction materials, industrial fodder) for private farms was not sufficient for the requirements of growth in production, and the bureaucratized state system for purchasing agricultural products was functioning deficiently. At the same time, the state monopoly of the market for basic agricultural goods (grain and meat) prevented any other alternative for farmers to sell their products. As a result, the development of agriculture began slowing down again in 1958.

A feeling of stagnation characterized Poland until 1970. This feeling accompanied the slowing down of the rate of economic growth, including agriculture; but the economy was not the only reason for it. Stagnation of political and cultural life contributed also.

It started in 1958 with the closing down of *Po Prostu* which was perhaps the most popular and most active of student weekly journals devoted to political debate. Student demonstrations followed that decision and these were forcibly suppressed. So-called "revisionists" were removed from all important positions within the party and state structure. The reign of persecution culminated in 1968 with the crushing of students' and intellectuals' protests and with "anti-Zionistic" campaigns; Polish citizens of Jewish descent who had escaped the Nazi holocaust during the war were evicted.

After 1958 all spontaneous social initiative was effectively suppressed. Worker-councils became an institution to control workers, supplementary in this function to other organizations already transformed into instruments of party control, such as workers' unions. The activity of agricultural circles also evaporated when their members observed that they had no influence on vital decisions concerning their affairs such as the purchase of tractors.

In socialized agriculture, as in industry, the total production was growing due to increased capital and labor inputs; but the efficiency of labor was stagnant. The rigid system of planning and management effectively weakened the incentives for greater efficiency, and more labor and modern technology, if it was introduced at all, were not improving the situation. Usually the new technology was incompatible with other, unchanged factors of production (employment, raw materials, etc.).

Although employment in services, including productive services for private agriculture, developed much faster than in the 1948-1956 period, neither quantity or quality of services was sufficient to satisfy the public demand. State-owned workshops were treated as an unimportant branch of the state economy and, therefore, were neglected in the supply of technical means and raw materials. General lack of incentives in the system, rigid planning and management, were obviously much more harmful for the functioning of services and production of consumptive goods than for heavy industry. Private workshops were tolerated but their access to technology and materials met more difficulties than state-owned industry. Under conditions of rigid price control and shortages of raw materials, private business could not be profitable; unless it engaged in illegal activity (buying stolen materials, charging prices higher than allowed by the state, etc.).

In agriculture the rise of production was also slow; in the period 1956-1960 the total agricultural production was rising about 3 percent annually; in the decade 1960-1970 about 2 percent annually. This rate of growth could not satisfy demand. Between 1956 and 1970 the urban population rose from 12 million to 17 million, the rural population remained constant (15.5 million), but the percentage of rural non-farm population was growing. A part of the government's demand for agricultural products, particularly for meat, was due to the rise of consumption aspirations. Perhaps one should note that such rising demands suggest an already eroded socialist ideology. The gap between demand for agricultural products and the level of agricultural production was growing; it was in this period that Poland started to import large quantities of grain.

The slow rise of agricultural production was caused by several factors. Apart from the lack of economic incentives for private farmers (the system of obligatory deliveries and low prices for agricultural products) and insufficient supplies of agricultural technology, demographic changes were also unfavorable for private agriculture. Constant migration of young people from the countryside resulted in an increase in the age of the farming population. In consequence, the number of farms without successors was steadily rising, as legal successors (young people who had migrated to cities) did not want to return to agriculture. Several regulations made it very difficult to sell land to other private farmers who had sufficient labor resources and were looking for acquisition of additional acreage, i.e., land from farms without successors and those where age of the producer made efficient farming impossible. As a consequence, much land was ceded to the state in exchange for a state pension (not being state employees, farmers were not granted state pensions). The percentage of land owned by state farms was growing and "socialist transformation" of Polish agriculture was advancing, but contrary to the postulate declared by the party, this transformation was not accompanied by increased productivity per hectare. State farms already were having problems with the fulfillment of their planned tasks and were not able to restore the productivity of these small and dispersed pieces of land taken over from private farmers.

Various measures to raise productivity in agriculture brought no improvement. By the late 1960s, the production of fertilizers was raised to a level which could roughly satisfy the demand (in 1970 the use of fertilizers was 124kg per hectare). But the fertilizers did not appear on the market in a manner which permitted farmers to buy them according to their needs. Instead, each farmer was obliged to buy state-determined amounts of fertilizers, independently of what was needed and these purchases usually were not delivered at the time when he needed them. In effect, farmers perceived this kind of distribution as a form of taxation. Another regulation was introduced ordering farmers to exchange a given amount of grain for "first quality seeds." But this seed was very often of much poorer quality than what the farmer could produce himself. Thus, a better *supply* of the means of agricultural production did not bring much improvement because the system of distribution was planned with the aim of getting administrative control over farmers rather than to help them produce more food.

The authorities were reluctant to supply farmers with modern technology even while imposing the administrative control. The decisive priority in the supply of all means of agricultural production has always been given to state farms and collective farms. Far behind them were members of agricultural circles, if they agreed to use jointly the technology provided to them, and still further behind were average farmers simply fulfilling the plans imposed on them by the system of obligatory deliveries or by contract production. In 1970 socialist agriculture owned 25 percent of the land but produced only 14.3 percent of the total agricultural output. State farms, collective farms, and also joint farming organized by agricultural circles produced less per hectare and had higher costs per unit of produce in general than private farmers. The price for the doctrine of socialist transformation of agriculture in Poland was inefficiency and low productivity of Polish agriculture and the growth of dependence on imported grain.

1970–1980

As the 1960s drew to a close, it was apparent Poland faced an acute economic disequilibrium which manifested itself in many ways but perhaps most publicly in the growing shortages of consumer goods. The roots of the problem were to be found in the increasing costs of operating the national economy, caused mostly by cancerous deficiencies in the centralized administrative system of planning and management. Also, there had been neglect of agriculture, as noted above, and an unbalanced industrialization program which was export-oriented and did not serve domestic consumer or agricultural needs.

To restore the equilibrium, or at least reduce the disequilibrium, the plan of raising consumer prices was adopted. Predictably, this led to a reduction of the living standards of the working classes. Violent strikes and riots began in 1970 among shipyard workers, spread to other branches of industry, and were prolonged until mid-1971. By the time the riots were bloodily suppressed, the political leadership had lost its support among the ruling elite and was removed.

In order to calm the situation, the new leadership raised wages and granted some privileges to workers in key industries, for instance, a better supply of cafeteria shops. These measures sharpened the disequilibrium even further. A solution then was sought in the use of foreign credit rather than in domestic economic reforms, which had been promised in negotiations with rioting workers. Use of foreign credit was a sharp break with the former leaders who had been trying to avoid this strategy for the modernization and stimulation of the economy. But the new leadership announced a very ambitious plan of economic development using foreign credit, particularly for intensification of export production. Some measures also were introduced to reactivate the private initiative in handicraft, services, retail and agriculture.

In agriculture the most spectacular measure was the abolition of obligatory deliveries which had the effect of encouraging the creation and growth of specialty farms. The freedom gained by the abolition provided more incentive for farmers, especially those with sufficient capital, to develop the most adequate and profitable branches of agricultural production. Specialty farmers also gained better access to the supply of agricultural technology. The lessening of restrictions for selling some agricultural products on the free market (eggs, fruit, vegetables), the rise of prices and benefits for some agricultural products within the system of contract production, and the abolition of restrictions for selling tractors and commercial feed to private farmers were also beneficial for the group of farmers who were able to undertake the risks and costs of investments.

Specialty farmers were directly assisted by the state through easier access to investment credit at the bank. In this way a privileged elite was created among the private farming sector. This select group of farmers was closely connected with local and national state bureaucracies and was strongly dependent on the continuation of state support through profitable contracts (often for export production), credits, supply of agricultural technology, construction materials, energy, etc. Because of this close collaborative relation, some party ideologists regarded specialty farmers as "more socialistic" or "more socialized," others denounced them as "rural capitalists." In fact, in a short time the standard of living of specialty farmers became conspicuously high.

But their role in agricultural production was not very impressive, particularly in supplying urban markets. They were not producing commodities which affected the supply of food for domestic mass consumption, such as meat, sugar, milk, potatoes, etc. Instead they were producing greenhouse flowers and early vegetables, fur animals, hogs for export and similar items of little consequence to the domestic food shortages.

The new stratification of Polish farmers, created by the agricultural policy of the early 1970s, was stimulated also by other acts of the state; two were particularly important. One was the introduction of retirement pensions for private farmers (1974). This act was seen by many as discriminatory since pensions of private, rural producers were set much below that of workers in other sectors of the economy, even if a farmer agreed to cede his land to the state. With the exception of the rural elite, most farmers were antagonized by the act and the first independent farmers unions were formed in response to it.

The second was the reform of regional administration which granted to the local, state-appointed, executives the right to undertake arbitrary measures in order to ensure fulfillment of agricultural production plans in private farming. In fact, the role of chief executive (*naczelnik*) in agricultural matters was comparable to that of the director in industry. "Naczelnik" was given the power to make decisions about the organization of production on private farms, to allocate profitable contracts and technology, to order consolidation of land and even to expropriate private land, if it was thought essential to the plans of development. This act put private farmers in the position of industrial workers, but without bargaining power.

In 1970 and 1971, when hopes for liberalization of agricultural policy were still strong, stimulated by initial acts of new leadership, and anti-peasant measures had not yet been undertaken, agricultural production increased at a very fast rate. In those two years, the rate of growth jumped to nearly 9 percent annually. This was due to better supplies of fertilizers (153kg NPK per hectare in 1974) and of modern technology (number of tractors rose in 1974 to 364,000). Even though the new technology was mostly allocated to state farms, this time private farmers also got some part of it.

The "great leap" of the economy didn't continue for long, however. Contrary to the promises made in 1970, the new leadership of the party failed to introduce economic reform and the sources of the disequilibrium in the system continued. Foreign credits only offered short-term relief. They were used mostly for new investments in heavy industry which was represented by the most influential pressure group in the political system. They also were used to purchase technology that was never installed. The new technology made Polish industry dependent on imported materials and components. With few exceptions, such as the automobile industry, the investments did not enable the economic system to meet the increased purchasing powers of consumers. They did not lead to increased exports to the highly competitive Western markets. And very soon the output of consumer goods and further purchase of raw materials and components had to be restricted because of financial difficulties in paying the foreign creditors.

Again, a solution was sought in raising prices for consumer goods. And again, the attempt failed, for the same reasons as in 1970. The brutal persecution of protesting workers gave rise to various opposition movements. Perhaps the most significant was the Committee for the Defense of Workers (KOR), headed by intellectuals of various political persuasions, which later played a very important role in the formation of free trade unions. Prices were raised again, several times, without announcement but in a more gradual way; in 1978-79 the rate of rising prices was about 20 percent annually. However, it proved to be an insufficient measure to restore the equilibrium in the market and to stop the deterioration of the economy. The rising burden of investments in heavy industry, which cost much more than planned and are mostly unfinished even today, created an acute crisis in the supply of energy and raw materials. By the late 1970s, some industries were closed down, the rest operated at 30-40 percent of capacity, and many investment projects were left unfinished.

The measures undertaken for the development of agriculture were ineffective for the same reason that plagued the general economy: lack of substantial reform of the economic system. Starting in 1975, national agricultural production became stagnant again, and in 1980 declined about 10 percent in relation to 1979. This stagnation and decline was due to the general economic and agricultural politics of the state in addition to the inefficiency of the economic system.

First there was the negligence of agriculture in general. State investments ("second industrialization") were concentrated in heavy industry, depriving other sectors of the economy of labor and capital, especially those producing consumer goods and agricultural technology. Investments in agriculture were even lower than in 1970.

Second, there was the continuous taking of private land. This consisted of appropriation of land without successors (children willing to continue farming), enforced purchases of land from private owners and of expropriation of private farms under various legal pretexts. Between 1970 and 1980, the share of privately owned land diminished from 75 to 68 percent of the total agricultural area; and correspondingly the share of state owned land rose from 25 to 32 percent. In this way the aim of furthering the "socialist transformation of agriculture" was executed through "statization" of agriculture, not collectivization. The process was accompanied by the enlargement of the average size of state farms, from 520 hectares in 1970 to 1,748 hectares in 1979, which meant greater centralization in decision making.

Third, priority in the access to all means of agricultural production was restored to state farms, and to a minor degree also collective farms and the "cooperatives" of the agricultural circles. This meant that even those scarce resources which were allocated to agriculture primarily went to socialist farming. For example, the supply of fertilizers for state farms amounted to 312 kg NPK per hectare, and for private farming only 150kg. Also, the supply of commercial feed was 740kg per hectare in state farms and only 355kg per hectare in private farming. Of the 573,000 tractors in use in 1979, 330,000 were in socialized farming and 243,000 in private farming.

Fourth, there was the inefficiency of all forms of socialist agriculture. The value of agricultural production per hectare on state farms was about 1/4 lower than on private farms; but the material costs of production were about 1/4 higher. Comparing input and output in state and private farming, economists warned that "each hectare used by state farms anihilates about 6,000 zlotys of the national income" (Manteuffel, 1980).

The collective farms and particularly the "cooperatives of agricultural circles" were even less efficient. The worst were the highly specialized "industrial farms" where the costs of production were extremely high. They were using only commercial feeds and required 8kg for the production of 1kg of meat while family farms used 2.8kg and state farms 4.2kg.

Fifth, there was the creation of a privileged group of specialty farmers which further deprived the average farmers of access to the means of modern agricultural production: tractors, energy, commercial feeds, fertilizers and construction materials.

Last, but not least, the reason for the collapse of private agriculture, together with the whole Polish economy, was anti-peasant legislation: discrimination in the funds for retirement, land succession, autocratic local administration, etc.

Conclusions

The fundamental reason for the collapse of the Polish economy was the over-centralized system of planning and management. The deficiences of such systems are largely known; many countries experienced that during World War II. Lacking the market's self-regulating mechanisms, the system is ineffective in meeting consumer needs and, therefore, is wasteful and anti-innovative. In all East European, Soviet block countries serious economic problems have been observed recently. However, they have not been as severe as in Poland.

There are several reasons why other socialist countries are coping better with economic problems than Poland. The Soviet Union has an enormously larger resource base. There have been special economic privileges granted in East Germany. Hungary has allowed the introduction of market mechanisms. Bulgaria has a basically non-industrial economy. There is less involvement in COMECON in Rumania. Lower industrial investments were required in an already industrialized Czechoslovakia.

Poland started an extremely ambitious program of industrialization. Western countries delivered modern technology on credit but the main consumers of the products manufactured in Poland were not Western, hard currency markets, nor even Polish consumers. Polish industry was oriented primarily toward delivering goods to the Soviet bloc. Paying Western suppliers in hard currency for technological components and raw materials required by this technology, Poland produced goods to sell in the East European markets for rubles. Apart from the wasteful use of foreign credits, this relation explains much of the present economic difficulties in Poland, particularly the collapse of the industrial sector.

Agricultural policy contributed also to the economic collapse. The continuous attempts at the socialist transformation of agriculture and the consequent, continuous anti-peasant policy ruined private farming without being able to replace it. Even if the socialist transformation had been finished, the situation would not be better. Manteuffel (1980) estimates that extremely high subventions to agriculture would be needed in such a case. But at least the agricultural policy would lead to less disruption.

But the crisis in Poland is not only economic. It is, above all, political. The immediate reason for the massive strikes in 1980 was the rise of food prices and the shortage of food on the market, as it has been before. But behind it was a general distrust in the government and disgust with its official propaganda of success, ridiculous not only for the audience but also for those directing it. In 1980, striking workers could not be pacified by promises or threats as in 1956, 1970 and 1976 when the government was unable to use military force. Workers in 1980 demanded social guarantees in the form of public control over governmental agencies.

For the first time in a communist state, workers won the right to form independent trade unions. In a very short time the independent union "Solidarity" unified the majority of Polish workers and employees. Supported by students (Independent Student Union) and farmers (Rural Solidarity), this movement became, in fact, a national force demanding substantial political and economic reforms, and, above all, rights for self-government in all internal and national affairs. Rural Solidarity demanded a guarantee for the stability of peasant property, equality of access to the supply of technical means for agricultural production, self-government for peasant-cooperates, and local councils dealing with the vital problems of rural families.

The demands of Polish workers, peasants, students, and intellectuals were not acceptable to the political leadership in the Soviet Union. When the communist party in Poland ceased to represent any political force (its ideological authority was eroded long before) power was exercised directly by the ultimate power-holders, the military and security police commanders. In December, 1981 a "state of war" was declared, Poland was placed under martial law, and the national resistance broken by massive use of military force.

The crisis is far from being resolved, however. The economic situation in 1982 is as bad as before and the political situation much worse. The sources of social conflicts and the roots of economic crisis cannot be eliminated without reaching some sort of compromise. After the introduction of martial law, the chances for that are much smaller than before.

The strategy of development adopted in socialist states reveals serious deficiencies. In all socialist countries the economic problems are becoming increasingly serious, the discontent more evident, and social conflicts more frequent. What happened in Poland could happen tomorrow in any other East European country.

The experience of the Polish crisis has a deeper resonance. It is a voice which questions all basic doctrinal communist assumptions about economic and social development. In agricultural strategy doubts are expressed about the advantages of large-scale farming over family farming. Polish agriculture collapsed not because the size of farms were small, or because small commodity producers were inefficient and anarchic. It collapsed because of the attempts to introduce large-scale, centrally governed farming, because of neglect of economic laws (market as self-regulating mechanism), and because of the anti-peasant policy of a monopolistic state. Large-scale farming can be advantageous but it requires a market-oriented economy to prove its advantages.

In Poland, even in very unfavorable circumstances, family farming proved to be more economically effective and more socially advantageous. Insistence on implementing doctrinal ideas created a state of permanent instability and ruined agriculture in Poland. The same applies to other countries where reorganization of agriculture, undertaken for political and doctrinal reasons, may bring the same results, if it has not done so already.

References

Banaji, J. 1976. Summary of selected parts of Kautsky's "The Agrarian Question," *Economy and Society* 5 (1):2-49. (There is no complete translation of Kautsky's book; a French edition was published in 1970 by Mespero (Paris) and an Italian edition by Feltrinelli (Milan) in 1971.

Konstantinov, Fedor Vital'evich. 1965. *Historical Materialism: A Marxist Sociology.* Moscow: Novosti Press Agency.

Lenin, V.I. 1921. Speech to the Third Congress of Comintern. Pp. 240-41 in *Collected Works.* Moscow: Progress Publishers, 1977.

Manteuffel, R. 1980. Socio-economic sectors in Polish Agriculture and their effectiveness, *Review of the Polish Academy of Sciences.* Warsaw: Polish Academy of Sciences.

Stalin, Joseph. 1940. *Dialectical and Historical Materialism.* New York: International Publishers.

————. 1952. *Economic Problems of Socialism in the U.S.S.R.* New York: International Publishers.

Tepicht, Jerzy. 1973. *Marxisme et Agriculture: Le Paysan Polanais.* Paris: A. Colin.

Part 3

Technology and Rural Life

Chapter 8

Agricultural Technology and Agrarian Community Organization

John W. Gartrell

In his presidential address to the Rural Sociological Society almost two decades ago, Gene Wilkening sought to direct our attention towards the study of process and structure in social collectives (Wilkening, 1964). He argued persuasively that this was the means to a fuller understanding of social change in rural society. I begin this chapter with an assessment of recent efforts to understand technology transfer in agriculture as it affects rural communities. This appraisal builds upon Wilkening's emphasis on specialization, integration and adaption by speculating on the implications of technology transfer for dependency, and the consequences of dependency for uneven development, community autonomy, and inequality.

We know remarkably little about the actual effects of international and national development on local communities. In order to stimulate future research, this chapter speculates about the kinds of propositions that might be formulated to understand the effects of technology transfer on rural communities.

Innovation

As Fliegel and Van Es have noted in chapter 1 of this volume, prior to the 1970s innovation research in rural sociology was dedicated to the proposition that technological improvements in agriculture were desirable. Innovation's consequences were rarely considered, and community characteristics were viewed as facilitating or inhibiting the diffusion of innovations. The reason for this positive evaluation of technological change was straightforward enough. Technical

149

improvements brought greatly increased farm production, a marked improvement in material existence, and benefits which clearly extended beyond the rural sector itself. The study of technological change in agriculture focused upon the communication of innovations and individual farmers' decisions to adopt new practices. Little attention was paid to the social process of invention itself, or to the consequences of trial and adoption of new behaviors (Goss, 1979; Rogers and Shoemaker, 1971). The classical diffusion model concentrated upon individual motivations and behaviors and ignored the consequences of technological change for the social collectives within which diffusion occurred.

Indeed, after reviewing studies of American rural communities, Sanders and Lewis concluded that:

> Although many of the studies [of factors associated with diffusion] have been located in a community and even bear the term "community" in the title, relatively few have tried to connect the findings to community variables. The emphasis is upon the characteristics of innovators, the change agents, or the stages of the adoption process. In other words, the community context is accepted as a given within which rural people conduct their lives (1976:40).

Technological change and rural communities appeared to be effectively linked only in studies of locating (or relocating) industrial plants or large scale energy resource projects in or near small rural communities. Even here, research has been comprised primarily of case studies that have not been conceptualized in terms of community differences (Summers, et al., 1976). Recently, researchers have shown an increasing concern for the effects of innovation in agriculture (Hightower, 1973; Rodefeld, et al., 1978; Newby, 1980; Singi, et al., 1981) and changes in community organization (Snipp and Summers, 1981; Bloomquist and Summers, 1982). Yet technology transfer and community change have seldom been linked in empirical research.

When the community was considered, it was often as a source of norms and values that influenced the perceived utilities of innovations (Flinn, 1970). Normative acceptance was seen as a determinant of who would innovate, low status deviants or high status elites. Studies of community social structure and agricultural innovation sometimes encountered severe methodological difficulties (Gartrell, 1976a) and ignored the consequences of innovations except to speculate about policy implications (Fliegel, et al., 1968; Fliegel, 1969; Gartrell, 1977; Gartrell and Gartrell, 1979). The only studies of community consequences of agricultural innovation took place in underdeveloped countries (Stanfield and Whiting, 1972; Havens, 1975; Havens and Flinn, 1975; Epstein, 1973; Gartrell, 1981). As Goss (1979) notes, this attention may have been drawn by the importance of easily identifiable rural communities (villages) in much of the Third World, by the failure of development programs, and by the rather obvious unequal distribution of benefits from change.

Modernization and Dependence

What clues can we draw from this experience that will inform our efforts at theory construction? American sociologists' efforts towards modernization of agriculture in low income nations met with only limited success. As authors and advisors, Schultz, Rostow, McClelland, Hagen, Moore, Hoselitz, Bendix, Lipsett and a long list of others all assumed that technological change was the road to economic growth and the pluralistic social structures that would free mankind from the double bonds of poverty and tradition (Rogers, 1976; Beltram, 1976). Neo-Marxist analysts have long disagreed. To them, the diffusion of technology was an imperialistic mechanism which supported the dominance of Asia, Africa and Latin America by western powers (Griffin, 1978; Amin, 1976; Frank, 1966, 1967). The development of capitalism was based upon the underdevelopment, impoverishment and structural distortion of colonial and tributary nations. Rural communities in underdeveloped nations suffered the double jeopardy of dominance by the local metropole as well as colonial, imperialist powers and multinational corporations. Modernization theory described underdeveloped nations in terms of dual or segmented societies. Dependency theory emphasized the asymmetry of exchange between metropolis and hinterland rather than the absence of effective rural-urban linkages. These relationships were described with the metaphor of internal colonialism (Hechter, 1975; Griffin, 1978; Portes, 1976).

The modernization approach to community development produced little systematic empirical research or codified propositions regarding the impact of technological change upon rural communities. Dependency theory and descriptions of internal colonialism have done little better. Both have focused upon the society as a whole rather than the rural community per se. What we have instead are distinct ideological differences as well as disagreements as to "fact" in characterizations of the relationships between hinterland and metropolis. On the one hand, dependency theory describes this relationship as one-sided, even predatory. Milder versions of such critiques refer to urban bias in rural planning. Against this are central place descriptions of rural growth centers, urbanization of the hinterland, and a more symmetrical "integration" of the local community into the national society. The "little tradition" influences the "big tradition."

Industrial technology originates outside the rural community and its diffusion is part of an asymmetric rural-urban relationship that creates uneven development and lowers community autonomy. After examining the effects of technology transfer on rural-urban relationships themselves, we will turn to the changes that are often brought with the introduction of "industrial" technology in agriculture: the changes in the division of labor, specialization and community heterogeneity, and class structure and inequality associated with the development of agrarian capitalism.

Rural-Urban Relationships

Looking from the bottom up, from the viewpoint of the rural community, the origin of technological change is primarily external to the community. Indeed, not only is the origin of new technologies external to the community in question, it is often foreign to the nation in which the community is located. The planning for change takes place in government offices and board rooms of international agencies and trans-national corporations (Warner, 1974). As Wilkening (1964) and others have noted, in the advanced industrial societies these forces have sounded the death knell for the small family farm and the autonomy of the rural community (Stein, 1964; Warren, 1978; Warner, 1974; Wilkinson, 1978; Bertrand, 1978). As other chapters in this volume have argued, the technology transferred in this process was not always appropriate to local institutions.

Technology transfer to rural communities is part of an asymmetric hinterland-metropolis relationship. As a consequence of this asymmetry, what is appropriate technology for metropole interests is not necessarily appropriate to local rural community interests. With the integration of the hinterland into the national society through this technology transfer, community autonomy is decreased. The more dependent the rural community becomes upon externally produced agrarian technology and national and international markets, the greater the focus of intercommunity relationships as ties to the metropolis. Network density in relationships between rural communities decreases. These speculative propositions suggest that as local markets decrease in importance, rural communities become alienated from one another even as they become more dependent upon metropolitan society. Interestingly, these propositions have an analogue in the international relationships between nations. Just as the dependent countries' exchanges within the world economic system tend to focus upon their metropoles, so too does the dependent rural communities' relationships to other social collectives within the nation focus more and more on asymmetric rural-urban exchanges and less and less on more symmetric exchanges with other rural communities. As is the consequence within the world system for nations, the concentration of ties within the nexus of asymmetric relationships with the metropole produces uneven development of communities.

Uneven Development

Although dependency theory has often viewed the hinterland as relatively undifferentiated, rural communities are not homogeneous. Differences in types of production systems and variation in the degree of penetration of rural communities influence the level of community development and variability in community social structure and process. Integration of communities into broader national collectives alters local community social structures. At the societal level, the degree of heterogeneity across rural communities will be a negative function of the level of development in the society as a whole. This proposition assumes an antecedent state in which rural communities are relatively isolated from each other and from the metroplis. It is not meant to imply that there is no extraction of surplus by the state, nor is it meant to ignore the fact that cities depend upon rural areas for food and other materials. The proposition is intended to reflect the influence of metropolitan dominance, and the increase of this dominance with the development of communications and transportation technologies in conjunction with complex organizations as a means of social control. It assumes that increasing potency of a relatively uniform source of influence upon rural communities will tend to make them more alike in the long run.

Differences in local history, ecology, resource bases and production systems result in considerable heterogeneity in local forms of social organization. Technology transfer within a country will be enhanced by the development of the nation state, which will in turn tend to reduce the effect of these differences. This will result in reduced heterogeneity across communities in their organizational properties. Technological change in communications and transportation and the development of national and international markets thus increases the influence of urban metropole and national and international elites. While the structure of these collectives is far from monolithic and elites anything but homogeneous, their long-run influence on different rural communities in urban-rural dominance relationships is similar in kind if different in quantity. Effects may thus differ in amount and timing from place to place, and the consequences for rural community organization may differ by type of production system (plantation, commercial farms, small holdings). Eventually, however, we would expect a consistency in urban-rural relationships to follow from consistencies in the source of dominance.

As a function of the timing of change, the development of capitalist agriculture may initially increase the heterogeneity across rural communities within a single nation. For example, the Indian government's efforts at redistribution (land reform, taxation) have been easily blunted by local elites who have captured the bulk of development benefits (Frankel, 1971; Griffin, 1974; Gartrell, 1976b; 1981). Government programs were concentrated in areas with complementary factors of production, particularly an assured water supply. Community differences in resource development were increased with a resulting increase in the unevenness of development.

In general, as capitalization, mechanization and commercialization of agriculture proceed, differences between communities in their level of development should decrease somewhat. We would therefore expect the degree of heterogeneity across rural communities within nations to first increase and then decrease as the level of (capitalist) development increases. Put somewhat differently, as the type of production systems in agriculture become more homogeneous throughout the country, differences between rural communities would be expected to decline.

A national economy as a whole may be experiencing increased specialization, and individual farms may increasingly specialize. Yet within a single rural community this may lead to greater homogeneity. The differences between communities at any one time would therefore reflect different types of production systems, different forms of capitalist development, and different degrees of penetration of agriculture by markets. The level of community development would therefore be expected to vary more within underdeveloped nations than it would within the nations at the core of the world capitalist system. Given the present state of development of that system, uneven development itself is perhaps a negative function of national development levels, and even underdeveloped nations may be on the "down side" of the curve.

To what degree is community "underdevelopment" a result of the dependence of local communities on the wider society? Is the accumulation of resources within the community and the kind of social structure it exhibits based upon more favorable resource endowments and higher levels of technology that depend upon rural-urban exchange? What conditions foster more beneficial effects of technology upon local social organization?

Community Dependence

Theoretically, the ratio of inputs to local social collectives garnered from within the community itself to those obtained from outside the community would indicate the level of input self-sufficiency. For example, seed may be generated from local production or bought from other areas. Villages may practice exogamy or endogamy in marriage markets. Labor may come from within the community, from neighboring villages, or from other regions entirely. Culture and recreation may come to the village in the "great tradition" or they may be generated locally within the "little tradition." Similarly, the level of output self-sufficiency would involve the export, reinvestment or local consumption of produce, people, ideas, and even legitimacy itself.

One difficulty in identifying the role of technology in precipitating community change lies precisely in the conception of community development levels as accumulated resources. New and old technologies facilitate resource accumulation to the point where it may be difficult to tell one from the other. In one sense technology is the relationship between production factors. In another, it is the level of development itself (Lenski, 1966). Certainly yesterday's resource inputs are today's accumulated resources. No matter what the level of societal development, contemporary technological change in agriculture often originates so thoroughly outside rural communities that current levels of both human and material resources may be difficult to distinguish from the very inputs that might indicate dependence.

For example, government material and personnel inputs constitute penetration of a rural community and also reflect, in part, that community's integration into the wider society. The community is dependent on the larger nation (governments and industries) for these inputs. Literacy or education, information on farming practices, machinery, fertilizers, pesticides, seeds, electricity and other energy resources (particularly inorganic ones), transportation and even water for irrigation all originate outside local communities. Technology and development levels are generally very highly and positively intercorrelated, and both are directly related to community dependence. However, in my sample of 84 agrarian communities in Andhra Pradesh, India, the level of development (farm capital per household) was virtually orthogonal to many of the indicators of dependence: the use of money in service relationships, the isolation of the village, and the degree to which service exchanges were located within the village (Gartrell, 1981). On the other hand, the level of capital per household was positively correlated with communications resource levels, electrification, contact with change agents, government agricultural inputs, and the sale of crops. The relationship between community development, technology, and dependence may not be as simple as it first appears.

Pre-capitalist peasant societies have generally been characterized as relatively self-sufficient, with little participation by villagers in the world outside their village and region (Wilkening, 1964). Production is labor extensive, with little time for exchange in markets for other goods and services. Social structure is kinship (family) centered, and labor relations are conducted on a personal basis. Even the spacial configuration of modern commercial farming communities has been altered by the "market demands" and product specialization that characterize commercial, industrialized agriculture. Farmers depend upon many specialized services in urban areas. Rather than a patriarchal village centered on attachment to land and integrated by common heritage, the modern commercial farming community is more loosely based upon the coincidence of residence reinforced by mutual interests and mutual help. The social structure is class centered and dominated increasingly by employer-employee relationships (Wilkening, 1964). Rural communities may even be composed primarily of laborers, with absentee owners living in urban areas. Under such conditions, rural community social structures become distorted, partial, homogeneous, and socially underdeveloped.

Community social organization is influenced by the spacial organization of production, the external origin of technological change in agriculture, and characteristics of the technology itself. To the degree that increased specialization and market dependence reduce community self-sufficiency and community autonomy, the traditional basis for integration is undermined. The conflicts that this may engender and the apathy and alienation it may bring reflect the "underdevelopment" of local communities. The external influence of national governments, international agencies, and national and international corporations may undermine local elites. The benefits of increased agricultural production may go primarily to capital whether it be domestic or foreign, private or state controlled.

Inequality

The benefits from increased productivity are usually not equally distributed. Property rights or other laws are likely to concentrate increases in the hands of an elite. Thus, the higher the level of community development, the greater the inequality in the distribution of capital and incomes within the agrarian community. There appears to be some evidence to support this proposition both in Brazil (Stanfield and Whiting, 1972;Thiesenhusen, chapter 13 in this volume) and in India (Gartrell, 1981). However, at higher levels of development, the increasing homogeneity of the division of labor within each community (despite specialization) may result in decreasing inequality. If the community is composed only of laborers, inequality is likely to be low.

In market economies, rural elites who benefit from development are generally the owners of land and other forms of capital. In non-market economies the elite are more likely to be the holders of power. In either circumstance, increased surpiuses in agriculture may be bled off by absentee, urban based elites, primarily interested in financing urban industrial development or simply their own conspicuous consumption.

Agrarian development programs which focus upon improvements in the quality of rural life may contribute to lower inequality. This appears to have been what happened in rural India, at least in some respects. Community literacy levels were negatively associated with inequality after controlling for the level of economic development and the dependence of the community on the wider society (Gartrell, 1981). Indeed, Adelman and Morris (1973) have suggested that such effects are observable throughout the Third World.

The characteristics of the technologies themselves can also influence their effects on local social structure. The divisibility of new technologies (economies of scale) coupled with the organization of agricultural production have a profound effect upon community social organization. Where new technologies promote more efficient production from existing resources, increased benefits go to those who control the new inputs and the "old" factors of production. For example, large scale irrigation works make other factors of production more effective and also serve as a means of social control for local elites or central governments. On the other hand, in Third World countries such as India, highly divisible technologies such as seeds and fertilizers may have distributive effects blunted by a lack of crop insurance or restrictions on markets for new products. And even when new technologies are highly divisible, increased benefits may not extend to labor (Frankel, 1971). In the U.S.A. and Canada, mechanization contributed to the concentration of capital in agriculture (Bertrand, 1978), and tractors and tube wells appear to have had similar effects in Asia (Gotsch, 1972).

Where technologies themselves do not so directly promote the concentration of capital, influences external to the rural community (supply systems, credit systems, marketing agencies, etc.) may do so anyway. Highly divisible technologies are often labor "intensive", requiring a large amount of labor per unit of investment in technology. Where labor is abundant and poorly organized, as is often the case in rural society, returns to labor are likely to be low and returns to other production factors high. Inequality is thereby exacerbated. Where new technologies radically alter the returns to various factors, or where change creates new opportunities outside the control of local, regional or national elites, technological change may have distributive effects (Epstein, 1973). At first glance, however, such situations appear to be rare.

From a different perspective, the effects of technological change in agriculture upon inequality within rural communities operates through migration. For example, mechanization in agriculture may "free" labor to move to cities by depriving laborers of employment. Per capita incomes within a community may improve and inequality be reduced if the disadvantaged poor are "forced" to leave the community. The same results occur, of course, even if pull factors are predominant. Migration effects upon levels of per capita incomes and inequality may be partially blunted if only part of the family moves, or if the move is temporary. On the other hand, to the degree that these new urban workers return part of their wages to the village, community income levels may be slightly augmented and inequality slightly reduced. The same effects would result from return migration if it is accompanied by savings used to generate incomes. This may occur, for example, through privileges given to those who serve the state (military).

If the relatively well-to-do or members of the local elite leave the community in order to enjoy a more modern lifestyle or to take advantage of opportunities (education, business, politics) available in urban areas, per capita incomes of the remaining community residents would be lowered. So too would inequality. If those who leave retain property within the village and continue to enjoy the rights of landlords even while they are absent, they may also retain considerable local political control. These absentee owners may be more inclined to import capitalist forms (mechanization, commercialization, etc.) into the village, but they may not want to decrease labor inputs. This might depend upon their interests in social control and politics. It might also depend upon the difficulty they experience with long distance management.

In India, members of the local village elite who migrate often maintain their residences in their home communities, reside there part of the year, and rely on some family member who lives in the community to manage their local interests. Increased surpluses from agricultural enterprises may be invested in urban production and often support high levels of consumption. Indeed, given the lack of taxation on agricultural production in Indian states such as Andhra Pradesh, and given the traditionally high value placed on owning property, investment in agricultural lands is widespread. At any rate, the urbanization of local elites and their integration into national bourgeoisie is often initiated by the education of their offspring in urban centers and their marriage into regional or urban elites.

In all of this, the real beneficiaries of technological development in agriculture are likely to be urban capitalists and (state) managers who want to keep food costs down so as to reduce the costs of urban labor. The benefits of technological development within agriculture may be captured by companies who manufacture machines, seeds, fertilizers, fuels and other factors in agricultural production. New profits may go to those who process, transport, wholesale, or retail farm produce. In market economies, new technology in agriculture is often created and developed by public institutions, but it often benefits private capitalists. Such developments are just as likely to benefit those who develop, produce, market and control the new technology as those who actually use it. There is no guarantee that benefits will be retained by rural communities, or even rural elites.

Integration

Technological change has occupied center stage in the theories of development economists, evolutionary anthropologists, and structural-functionalists of all disciplines. Indeed, only with the development of inorganic sources of energy did control of reproduceability of technology shift beyond the boundaries of rural communities and local exchange systems. According to these theories, technological development allowed the size of social collectives to increase and brought increasing social differentiation. The division of labor became more complex, as did the regulatory organizations necessary to coordinate activities. In these theories the principal consequence of technological change was increased differentiation and the main problem was one of social integration.

Does technological change in agriculture have similar effects upon rural communities? Interestingly, the results may be the same, but the reasons may be different. The urbanization, bureaucratization, and centralization of society which is involved in the development of market or state capitalism creates political emasculation, lost autonomy, and high out-migration which in turn result in increased problems of integration within rural communities. If the antecedent is increased differentiation, its effects are indirect. Technological advance may not bring population increases to rural communities but rather population declines. Certainly the mechanization and commercialization of agriculture drastically alters roles associated with the occupation of farmer. Yet the specialized occupations created by local demand are primarily located outside the rural farm community.

Some rural communities may disappear, particularly those whose functions are made obsolete by rural growth centers or changes in trade and transport. This appears to be characteristic of more advanced industrial societies. Yet even in vastly different circumstances, technological change may produce a decrease in heterogeneity within rural communities. Rural communities experience a decrease in heterogeneity with the decline of handicraft and cottage industries associated with industrial development. As capitalist development proceeds, the spacial concentration of transport, marketing, governance and most agricultural inputs and services follows the concentration of capital. The establishment of public sector agencies, industrial manufacturing firms, resource extraction enterprises, or energy production industries in or near rural communities may increase differentiation. But these changes result from developments outside of agriculture, and even here, the benefits of change have been elusive to local communities (Summers, *et al.,* 1976).

Again, it is interesting to note the parallels in agricultural systems to the results predicted in dependency and world systems theories for nations as a whole. At the community level the conditions of capitalist development do not necessarily produce the differentiation that is assumed to be the outcome of increased levels of production. The differentiation that is produced may not be located within rural communities, and it may not be the basis for reductions in inequality, or changes in integration. The strong social cohesion of the peasant community may be replaced by an organic solidarity that binds agrarian communities to the wider society. However, within the local community itself, integration may be weakened, particularly with high immigration. The ties that bind become more exclusively the ties of common residence, and common position in the wider society (Wilkinson, 1978; McGranahan, chapter 9 in this volume).

Conclusion

A reappraisal of efforts to understand the effects of technology transfer in agriculture on rural communities has revealed how little writing and research has been focused upon the community as a unit of analysis. The preliminary attempts at theory construction outlined above suggest that because technology originates outside local communities, asymmetric rural-urban relationships provide one focus for such studies. Given this asymmetry, what is appropriate technology for metropole interests may not be appropriate to agrarian community interests. The greater the technological development level of the community, the lower its autonomy. As technological and other ties to urban centers increase, network density in relationships between rural communities decrease.

At the societal level, given integration into the world capitalist system, the degree of heterogeneity across communities in the level of development will be a negative function of the level of societal development. The development of capitalist agriculture may initially increase the unevenness of development across rural communities within a nation but as development proceeds heterogeneity will decrease. Increasing specialization on the part of individual farm operations and specialization within the economy as a whole will actually result in greater homogeneity within many rural communities. The differentiation that accompanies specialization is often located outside the rural community (in large urban and regional growth centers).

The higher the level of technological development within the rural community and the lower the level of community autonomy, the greater the community's dependence on the wider society. At lower levels of societal development, the higher the level of community development, the greater the inequality within the community. However, among advanced industrial societies at the core of the world system, higher levels of technological development may create lower levels of inequality within the community. Finally, increased dependence and lower community autonomy undermines traditional integration within the rural community as well as integration between rural communities.

Our understanding of the effects of technological change in agriculture upon rural communities requires a conceptualization of the community as a social collective, perhaps adapting models used for other social collectives such as societies or complex organizations. Wilkening's conclusion to his presidential address to the Rural Sociological Society almost two decades ago remains relevant today. "We must study change not only as a matter of individual choice and action, but also as a function of social systems" (1964:17). Methodologically he encouraged comparative research. Substantively, he urged a focus on social and other cleavages, the structure of power and influence, the extent and nature of social contact and communication, and the extent and nature of linkages between systems and sub-systems.

References

Adelman, Irma, and Cynthia Taft Morris. 1973. *Economic Growth and Social Equity in Developing Countries*. Stanford: Stanford University Press.

Amin, Samir. 1976. *Unequal Development*. New York: Monthly Review Press.

Beltram, Luis Ramiro. 1976. "Alien premises, objects, and methods in Latin American communication research." *Communications Research* 3(April):107–134.

Bertrand, Alvin L. 1978. "Rural social organizational implications of technology and industry." Pp. 75–90 in Thomas R. Ford (ed.), *Rural U.S.A.: Persistence and Change*. Ames, IA: Iowa State University Press.

Bloomquist, Leonard E., and Gene Summers. 1982. "Organization of production and community income distributions." *American Sociological Review* 47(June):325–338.

Epstein, T.S. 1973. *South India: Yesterday, Today, and Tomorrow.* London: MacMillan.

Fliegel, Frederick C. 1969. "Community organization and acceptance of change in rural India." *Rural Sociology* 34(June):167-181.

Fliegel, Frederick C.; Roy Prodipto, Lalit K. Sen; and Joseph Kivlin. 1968. *Agricultural Innovations in Indian Villages*. Hyderabad: National Institute of Community Development.

Flinn, William L. 1970. "Influence of community values on innovativeness." *American Journal of Sociology* 75(May):983-991.

Frank, Andre Gundar. 1966. "The development of underdevelopment." *Monthly Review* 18(September):17-31.

_____. 1967. "Sociology of development and underdevelopment of sociology." *Catalyst* 3:1-67.

Frankel, Francine R. 1971. *India's Green Revolution: Economic Gains and Political Costs*. Princeton: Princeton University Press.

Gartrell, John W. 1976a. "Comment on Village Influence on Panjabi Farm Modernization." *American Journal of Sociology* 81(March):1169-1173.

_____. 1976b. "Inputs and Inequality." *Kerala Sociologist* 4:152-166.

_____. 1977. "Status, inequality and innovation: the green revolution in Andhra Pradesh, India" *American Sociologial Review* 42(April):318-337.

_____. 1981. "Inequality within rural communities of India." *American Sociological Review* 46(December): 768-782.

_____, and C. David Gartrell. 1979. "Status, knowledge and innovation: risk and uncertainty in agrarian India." *Rural Sociology* 44(Spring):73-94.

Griffin, Keith. 1974. *The Political Economy of Agrarian Change: an Essay on the Green Revolution*. London: MacMillan.

_____. 1978. *International Inequality and National Poverty*. New York: Holmes and Meier.

Goss, Kevin F. 1979. "Consequences of diffusion of innovations." *Rural Sociology* 44(Winter):754-772.

Gotsch, Carl H. 1972. "Technological change and the distribution of income in rural areas." *American Journal of Agricultural Economics* 54(May):326-41.

Hightower, Jim. 1973. *Hard Tomatoes, Hard Times*. Cambridge, Mass.: Schenkman.

Havens, A. Eugene. 1975. "Diffusion of new seed varieties and its consequences: a Colombian case." Pp. 93-111 in A.E. Havens (ed.) *Problems in Rural Development: Case Studies and Multidisciplinary Perspectives*. Leiden, Holland: E.J. Brill.

_____, and William L. Flinn. 1975. "Green Revolution technology and community development: the limits of action programs." *Economic Development and Cultural Change* 23:469-481.

Hechter, Michael. 1975. *Internal Colonialism: The Celtic Fringe in British National Development, 1536-1966*. London: Routledge and Kegan Paul.

Lenski, Gerhard. 1966. *Power and Privilege*. New York: McGraw-Hill.

Newby, Howard. 1980. "Rural sociology—a trend report." *Current Sociology* 28(Spring) 1-141.

Portes, Alejandro. 1976. "On the sociology of national development: theories and issues." *American Journal of Sociology* 82(July):55-85.

Rodefeld, Richard D.; Jan Flora; Donald Voth; Isao Fujimoto; and James Converse (eds.). 1978. *Change in Rural America: Causes, Consequences and Alternatives*. St. Louis: C.V. Mosby.

Rogers, Everett M. (ed.). 1976. "Communications and development: critical perspectives." *Communications Research* 3(April).

_____. and F.F. Shoemaker. 1971. *Communication of Innovations: A Cross-cultural Approach.* New York: Free Press.

Sanders, Irwin T., and Gordon F. Lewis. 1976. "Rural community studies in the United States: a decade in review." Pp. 35–53 in Alex Inkeles (ed.), *Annual Review of Sociology* 1976 (vol. 2). Palo Alto, CA: Annual Reviews.

Singi, Prakash M.; Frederick C. Fliegel; and Joseph E. Kivlin. 1981. "Agricultural technology and the issue of unequal distribution of rewards: an Indian case study." *Rural Sociology* 46(Fall):430–445.

Snipp, C. Matthew, and Gene F. Summers. 1981. "The welfare state in the community: a general model and empirical assessment." *Rural Sociology* 46(Winter):532–607.

Stanfield, David J., and Gordon C. Whiting. 1972. "Economic strata and opportunity structure as determinants of innovativeness and productivity in rural Brazil." *Rural Sociology* 37(September): 401–416.

Stein, Maurice. 1964. *The Eclipse of Community.* New York: Harper Row.

Summers, Gene F.; Sharon D. Evans; Jon Minkoff; Frank Clemente; and E.M. Beck. 1976. *Industrial Invasion of Nonmetropolitan America.* New York: Praeger.

Warner, W. Keith. 1974. "Rural society in a post-industrial age." *Rural Sociology* 39(Fall):306–317.

Warren, Roland L. 1978. *The Community in America,* 3rd edition. Chicago: Rand McNally.

Wilkinson, Kenneth P. 1978. "Rural community change." Pp. 115–125 in Thomas R. Ford (ed.), *Rural U.S.A.: Persistence and Change.* Ames, IA: Iowa State University Press.

Wilkening, Eugene A. 1964. "Some perspectives on change in rural societies." *Rural Sociology* 29(March):1–17.

Chapter 9

Changes in the Social and Spatial Structure of the Rural Community

David A. McGranahan

Introduction

Contemporary American rural society bears a limited resemblance to the rural society that existed at the beginning of the century. By definition, rural society still involves people living in small towns and open country areas. But rural life is no longer organized primarily around agriculture as it once was. In 1970, only 14 percent of the labor force resident in nonmetropolitan rural areas was directly engaged in agriculture (U.S. Bureau of the Census, 1972). In addition, distances have shrunk with improvements in transportation, and rural life has become much less circumscribed geographically. This paper investigates the effects of these changes on the social and spatial structure of rural communities.

Rural community change is not a new topic but it has been treated from a largely functionalist perspective. Warren (1978), Wilkinson (1979) and others point to the increasing scale of local organization, its differentiation, its fragmentation and its orientation to the institutions of the broader society, all of which have resulted in problems of community integration. As insightful as these analyses are, they are nonetheless limited by their theory. Inherent in this and much other contemporary community research is the constraining presumption that common interests arise from common geographic location. As a consequence, rural class and other structural divisions are largely ignored, as are the differing geographic patterns of various rural activities. One result is that, given that our rural society is no longer a largely agricultural society, and that population size and density are now less constraining on the scale of local organization, it has become unclear if rural and urban communities continue to differ in sociologically important ways (see Newby, 1980; Newby and Buttel, 1980).

The perspective taken here is that an understanding of the rural community requires an analysis of the interrelationships between the social and the spatial structures of rural society. Local interests arise to the extent that well-being is affected by what happens in a geographic locale, but the meaning of geographic locale for well-being varies depending on a person's social role or activity. For instance, for a merchant, a town is a central place with a local trade area, while for a manufacturer it may be primarily a location with access to low cost labor. Since change benefitting one need not benefit the other, they have different local interests. The meaning of geographic locale may also differ with respect to its boundaries. For a wage earner, the locale is primarily the labor market area which, because of commuting, may now extend well beyond a town's trade area. On the other hand, social activities associated with residence may be largely confined to neighborhoods, which are often smaller than business trade areas. Thus interests must be differentiated according to both social content and spatial scope.

Community coalition from this perspective may result from common interests arising from common location, but location is defined on both a social and geographic dimension. Different economic and residential groups may have separate interests and organizations with different territorial bases—neighborhood, trade area, labor market area and so forth. The community is not considered here to be a single entity but a spatially unbounded complex of coalitions and organizations whose social and territorial bases change over time as a reflection of changes in the social and spatial structure of rural activities and opportunities.

A Typology of Rural Activities and Interests

As a first step in differentiating rural activities and interests, it seems useful to distinguish those relating to income and economic opportunities (economic) from those relating to residence and consumption (residential). People with quite disparate economic roles and interests (e.g. merchants, salaried professionals) may have similar residential activities and interests. Moreover, because of absentee ownership people may have, as business owners, economic interests attached to a locale in which they have no residential role or interest.

Local economic interests are generated to the extent that income opportunities are determined by local market conditions, with different markets (e.g. labor, product) and positions in markets (buyer or seller) yielding different local interests.[1] Drawing largely on an earlier study of the spatial structure of rural income distribution, which showed how the effects of geographic location on income vary by economic position (McGranahan, 1980a), five economic groups are considered here: 1) farmers, who once sold in local and regional markets, but whose incomes are now largely determined at the national level; 2) merchants and owners of other establishments such as banks and utilities which operate in locally defined trade areas; 3) manufacturers and other employers who sell products outside the local trade area but who purchase labor in local labor markets; 4) managers and other highly skilled workers, who participate in increasingly broad labor markets; and 5) less skilled workers, who participate in local labor markets.

Residential interests stem from the social and consumption activities of households and the spatial organization of these activities. Patterns of social interaction within an area may generate a sense of community which is valued in itself by participating residents. The "community of sentiment" (Suttles and Janowitz, 1979), which provides members feelings of belonging and identity, is an important aspect of residential life and interests. While economic interests may be identified from positions in and boundaries of markets, residential interests are largely based on nonmarket activities and are less easily specified.

The analysis of rural community change is carried out below in three parts. Drawing largely on central place theory, which describes the spatial organization of markets in agricultural areas (Christaller, 1966; Smith, 1972), the analysis begins with a discussion of early (circa 1900) agricultural communities and the changing interests of farmers and central place business owners. The second section focuses on manufacturers, workers and rural labor markets, and the third discusses changes in the spatial patterns of residential activities and interests. Finally, contemporary rural-urban differences in the composition and spatial structure of activities and interests are noted. Although rural communities are no longer primarily agricultural communities, important differences between rural and urban communities remain.

Farmers, Merchants and Central Places

Farmers. Rural society was once a primarily agricultural society, ecologically organized into agricultural centers with surrounding farm populations. Writing in the early part of this century, Veblen described the farmer's economic relationship with the local center as follows:

> The country town is an organization of business concerns engaged in buying things from farmers in order to sell at an advance to the central markets, and in buying from the central markets in order to sell at an advance to farmers. The country town is an organization of middlemen and it is out of these differences between the buying price and the selling price that the entire town gets its living. (Quoted in Lipset, 1968:48).

In early agricultural communities, local economic interests for farmers were generated by the territorial organization of markets, which often permitted town businessmen local monopolies. Before the development of all-weather roads and the widespread use of the automobile and truck, relationships among centers appear to have been largely treelike or "dendritic" (Smith, 1972), in that towns were vertically linked to larger places in the central place hierarchy, but often isolated from each other. The lack of intercenter competition permitted local town monopolies as well as monopoly pricing at the wholesale level. Moreover, because products were shipped largely by rail, farm incomes suffered from monopolies over transportation links as well.

The local and broader based economic interests generated by this marketing structure were reflected in a series of farm movements during the late 19th and early 20th centuries. The Grange, and later the Farmers Alliance and Farmers Union, attempted to overcome the entire dendritic marketing system by promoting cooperative marketing structures at both the local and wholesale level and by promoting legislation to control railroad freight prices. These farm organizations were explicitly anti-town and anti-monopoly. For instance, local affiliates of the Farmers Union, started in 1902, were open to local non-farmers, but people in "banking, merchandising, practicing law or belonging to any trust or combine" were explicitly excluded (Quoted in Weist, 1923:481).

The economic divisions between local central place business owners and farmers were reenforced by a segmentation of social activities between town and country. While villages, in addition to serving as market places, were centers of social and political[2] activity for town residents, the primary social unit for many early farm families appears to have been the rural neighborhood. In their national study of 140 agricultural communities, Brunner, et al. (1927) found that the neighborhood as a social entity was not ubiquitous, but arose under various conditions such as ethnic group settlement, geographic isolation, and inmigration as familial or friendship groups, and tended to cohere around schools, churches and other institutions such as the Grange. Although this study defined community boundaries partly on the basis of social activities (trade areas tended to be larger than "communities"), between one-fifth and one-third of the open country community residents lived in coherent neighborhoods, depending on the region.[3]

As locales for the social activities of farm families and, at least before mechanization became widespread, as bases of work exchange among farmers (Kolb, 1957), rural neighborhoods were important in the mobilization of rural public opinion (Brunner, et al., 1927). In fact, the Grange was originally promoted as a neighborhood association patterned after lodges, with social and educational purposes, but quickly turned into an economic interest group organization at the local and state levels (Weist, 1923).

In sum, while relations between town and country, merchant and farmer, were undoubtedly smoother in some areas than others, our tendency to think of the early rural community with its service center and surrounding farm population as a social whole, as a basis of community identity and attachment, denies the important economic and social divisions of rural society at the turn of the century.

Since World War I, the economic importance of locality for farmers has declined considerably. Vehicle and road improvements, culminating with the construction of the interstate highway system, have extended the possible range of trade areas, reducing the isolation of individual centers and creating the more competitive situation envisioned in Christaller's (1966) network theory of central place systems. The development of cooperative marketing structures, increased state and federal involvement in price supports and, more recently, the considerable international marketing of farm produce have meant that the economic interests of farmers have become increasingly national in scope. Over time, the original bases of local economic conflict between farmers and local central place businessmen have largely disappeared. These two groups, in fact, have common interests at the state and national level, for high farm incomes mean higher incomes for local businessmen as well.

County level farm organizations persist in many areas. Agricultural extension services are organized at that level, and county government affects farmers both as business owners and as residents. However, reflecting in part the importance of broader level markets and policies, power in farm organizations is usually highly concentrated in the state and national level associations. (In contrast, merchant associations such as the chamber of commerce maintain considerable local autonomy.)

While this discussion has focused on farm product markets, farmers may also have local labor market interests. Farm families, especially those with marginal farms may desire opportunities for off-farm work, and in this respect be in the same position as less-skilled workers. Owners with large, labor intensive enterprises are in the position of employers with extra-local markets, with an interest in minimizing change that would raise local wage rates. The positions of workers and employers are discussed below.

Merchants and Other Central Place Business Owners. While transportation improvements have contributed to the decline in locality-based farm interests, for the owners of businesses serving local trade areas, the economic importance of locality has in many respects increased. The incomes and wealth of these businessmen depend not only on price margins but also on their volume of business, particularly where significant economies of scale are involved. For many central place businesses, such as banks, newspapers, construction firms and dealerships, the size of the enterprise is limited by the volume of trade carried out in the center in which they are located, by the center's commercial importance (see Hawley, 1941; Johnson, 1982). Center growth is necessary for business growth, and center decline means business decline. Differentiation within centers creates common center-based interests, and as the geographic isolation of centers has diminished, the similarity of establishments across centers has increased competition among centers. In general, intercenter competition has tended to favor larger places (Clements, 1978; McGranahan, 1980b). Larger centers offer not only a wider variety of services, but, because of economies of scale, their businesses are able to sell at lower prices. Center growth thus has become important not only in itself but because it enhances the competitive position within a network of central places.

This is not to argue that merchants and others with local trade areas have no regional or larger interests in higher farm prices or other changes which promote growth in the region. But the spatial pattern of growth (or decline) in a region is a crucial question for these businessmen and the center is of primary concern. The businessmen of a given center may find that they are losing trade if a neighboring center grows and gains in commercial importance.

For an agricultural service center, opportunities for growth are limited by the carrying capacity of the surrounding farmland and the debt capacity of the farmers. An alternative which has long been recognized by central place business owners is to promote manufacturing. Brunner, *et al.* (1927) in their nationwide study of 140 agricultural villages found that:

> The limitation of opportunities for employment. . .gives rise to the "payroll" fixation encountered by the field investigators in all parts of the country. The village business men are convinced that if factories could be induced to come in, with their weekly pay envelopes, they would not only stimulate trade but would increase the occupational opportunities of the town. Accordingly chambers of commerce or other similar groups are organized. Such organizations are reported in [83 of the 140 study villages].
>
> Strenuous and costly efforts are made to attract industries. Many examples are cited by the field investigators of the offering of bonuses, of free sites and of local capital optimistically invested in schemes to convert this village or that into a prosperous manufacturing center. (1927:126–127).

The opportunities to attract manufacturing appear to have been severely limited during this period, however, and there was little if any industrial expansion in these 140 villages between 1924 and 1930 (Brunner and Kolb, 1933). Much of the industry that did exist depended on local resources or farm output. Beginning in the 1960's, in part due to the development of the interstate highway system, rural or at least nonmetropolitan manufacturing employment began to increase substantially (Summers, *et al.*, 1976; Haren and Holling, 1979). Many of the new plants, not based on processing agricultural or other local resources, had relatively high locational flexibility. In this context, smaller centers, like major cities, became "growth machines" (Molotch, 1976) for local merchant interest groups.[4]

Changes in residential interests and their spatial structures have yet to be discussed. One of the characteristics of residence in the early agriculture communities, however, was the segregation of farmers and merchants. Since that time, labor has become an increasingly important class in rural regions. Residential activities, interests and segregation cannot be discussed without taking their presence into account. Before turning to residential interests, then, an analysis of employers, workers and labor markets is in order.

Employers, Workers and Labor Markets

Employers with Extra-local Sales. Manufacturers and other employers such as the owners of large scale farms and mining firms who sell outside of the local trade area have local economic interests which are quite distinct from and at times in conflict with merchants' local interests. For employers with extra-local sales, income may depend on a number of factors, including access to resources, to markets and to low cost labor, but center size and growth is itself of little economic benefit, and may, if it is associated with a rise in local wages, be against employer interests.

In early industrial, mining and large farm areas, workers usually lived and worked in the same place, and relations between local owners and residents were often an extension of relations between owners and workers. Sometimes the owners would completely dominate residential life through the ownership of housing, retail stores and the provision of health and other services (see, e.g., Caudill, 1963). Other times, relationships were socially paternalistic, with the owners participating in the residential community through public and charitable works (see French, 1969; Warner and Low, 1946). Contributing to and becoming part of the community of sentiment undoubtedly yielded paternalistic owners better worker relations within economic enterprises, although the purpose may not always have been seen from that light by either owners or workers (see Newby, *et al.*, 1979).

This extension of owner-worker relations into the locality has tended to decline over time for several related reasons. First, absentee-ownership, which has increased substantially, makes it difficult for employers to take an active, personalized role in the local setting. Second, industry-wide union contract negotiations in some of the capital intensive resource-based industries such as mining and paper has meant that labor costs in these establishments are determined independently of the local setting. Third, many of the new manufacturing establishments are labor intensive and do not, aside from labor, depend on local inputs. Their ease of mobility can act as a deterrent against local labor demands. Finally, because of commuting, place of residence is no longer the place of work for many people. Even a fairly large plant can move to a small town without becoming the predominant employer for the town's residents. On the one hand, this has meant that employer contributions to residential welfare are less likely to influence relations with workers. On the other hand, it has probably made concerted worker demands less likely, since the sense of community arising from common place of residence no longer reinforces any sense of community which might arise from working in the same establishment.

It is possible to overstate the withdrawal of employers from participation in residential affairs, however. A recent study in northwestern Wisconsin found that about a third of the manufacturing plants employing at least 20 people were locally owned, and about a third of the local owners had held public office. Where owners had held public office, none of their plants were unionized, while otherwise, nearly 40 percent of locally owned plants had unions (McGranahan, 1982). In this rural region, at least, employer-worker and employer-community relations retain some interdependence.

Perhaps the central point to be made here is that although the owners of both central place businesses and industries with extra-local markets may have had, as residents, the same local status, their local economic interests have always been quite distinct, and their local participation has taken quite different forms. Employers selling outside the local trade area do not form part of the local growth machine, and may, if growth threatens to raise area wages, come into conflict with local central place businessmen.

Workers. In early communities, workers and potential workers had local interests consistent with those of the merchants, at least insofar as center growth was concerned. Preferably, growth would result in higher local wages, but it at least reduced the need to migrate for those not able to immediately acquire farms or small businesses. Over time, two trends have reduced worker interest in central place growth.

First, bureaucratization and professionalization in the expanding nonagricultural private and public sectors has generated a large group of *highly skilled workers,* managers and professionals for whom economic advancement comes not in the local labor market but through professional associations and contacts and bureaucratic hierarchies which are regional and often national in scope. For this group, which has high residential mobility and, compared to less skilled workers, relatively equal incomes across areas (Chiswick, 1974; McGranahan, 1980a), local economic interests are weak.

Second, transportation improvements have made it possible for workers to commute increasingly long distances to work. The incomes of *less skilled workers* still depend on local labor market conditions, but, because these local labor markets have expanded considerably in geographic scope, the local interests of this group are no longer confined to the place of residence and its immediate vicinity. Thus, in contrast to the situation of central place business owners, for whom the growth of a nearby place may pose an economic threat, the economic interests of less skilled workers have become areawide.

These interests are generally not articulated through worker organizations. Rural workers are represented, if at all, at the plant level, and where union locals do exist, they affiliate extralocally by occupation rather than by geographic labor market areas. As a consequence, in contrast to the central place businessmen, less skilled workers have had little if any voice in the type and spatial organization of rural economic change.

In sum, the local town or central place was a focus of interest for most economic groups in early rural society, but, over time, these groups have been affected in quite different ways. For farmers and highly skilled workers, locality has lost much of its economic importance. The labor market area has replaced the immediate center for less skilled workers. However, owners of businesses serving local trade areas retain center-based growth related interests. With less skilled workers the major exception, the spatial bases of economic interest group associations generally reflect market structures. Farmers have county and multi-county associations with little local autonomy; professionals and other highly skilled workers have state and national associations; and those with local trade areas have chambers of commerce, industrial development corporations and other center-based organizations. Less skilled workers, however, tend not to be organized by labor market area.

Residents, Communities of Sentiment and Service Areas

Restudies of the 140 agricultural communities studied by Brunner in the early 1920's found that villages were absorbing the social activities of nearby rural neighborhoods (Brunner and Kolb, 1933; Brunner and Lorge, 1937). While they suggest that road improvements were making centers more accessible to open country residents, the process was undoubtedly facilitated by the declining economic antagonism between farmers and merchants. During this period, perhaps because of the depression, the number of stores in small towns was generally increasing, which led some observers to conclude that the village would eventually become the focal point of both social activities and economic services (e.g., Sanderson, 1939). But since the 1930's, villages have been losing retail and other functions to larger centers (Johansen and Fuguitt, 1973). People now use a hierarchy of service centers, making small purchases locally but travelling considerable distances to larger centers for major purchases. In contrast, social activities appear to have remained spatially more circumscribed (Kolb and Day, 1950; McGranahan, 1980b). Thus, although both have expanded geographically, the community of sentiment is, as in the early agricultural community, smaller than the area in which people travel for services (Haga and Folse, 1971; Clemente, Rojek and Beck, 1974).

Perhaps the major change that has occurred with respect to residential activities and interests is their loss of geographic association with economic interests. The division in the early agricultural community between country neighborhood and town was also a division between farmer and merchant. Farm and central place business associations could represent at once local residential and economic interests. At present, because of the increasing numerical importance of workers as a rural economic group, and worker residence in both town and country areas, local associations based on business and farm interests can no longer be said to represent local residential groups.

Few rural voluntary associations in rural areas are explicitly organized to represent local residential interests. Expressive associations such as social clubs may serve as a means of mobilizing public opinion, at least on those issues over which there is public debate. Local government itself may serve as a vehicle for articulating residential interests. However, given that central place businessmen are usually highly organized at the municipal level and that farmers have retained county level organizations, the extent to which local government actually represents residential interests when they conflict with the interests of business owners or farmers is unclear.

Contemporary Differences Between Rural and Urban Communities

With rural society no longer organized largely around agriculture, and rural life no longer confined to the personalized relationships of the neighborhood and local town, the question arises as to whether the rural community remains sociologically distinct from the urban community or whether it has become simply a smaller, more spatially diffuse version of the urban community (see Newby and Buttel, 1980). This question may be addressed on two bases, following the analytic perspective taken here. One is social, relating to the nature of rural and urban activities and interests, and the other is spatial, relating to the geographic organization of activities. Contemporary rural-urban communities appear to differ on both accounts.

With respect to the social aspect, there are certainly similarities in the activities and interests associated with contemporary rural and urban communities. Urban like rural residents have communities of sentiment. Owners of urban central place businesses, like their rural counterparts, benefit from central place growth. Highly skilled workers such as professionals and managers tend, irrespective of location, to participate in national or at least regional labor markets. There appear to be major rural-urban differences, however, in the activities and interests of employers with extra-local sales. Rural employers, it was argued above, hire in local labor markets. Their well-being depends in part on keeping area wages low. Major urban employers are in a different market situation, and have different local interests.

This difference can perhaps best be understood by reference to the dual economy literature, for the distinction made there between core and periphery firms and industries has a geographic correspondence.[5] To the extent that a firm (or plant) has what Baron and Bielby (1980) suggest are core characteristics — large size, high skill requirements, capital intensity, strong unions and internal

labor markets – access to skilled labor, specialized services and information may
be important, but area wage levels are of little concern. In contrast, periphery-
type firms, with smaller plants, routinized, labor-intensive production processes,
low skill requirements and weak unionization, have little need for access to specialized knowledge and information, but are affected by local wage rates. Rural
areas, with relatively low wages, dispersed population, and, at least when agricultural employment was declining, an excess of less-skilled workers have generally been more suitable for the latter type of firm, with the result that, as shown in
Table 9.1, highly urbanized locations have mostly core sector manufacturing,

TABLE 9.1

**Percentage Distribution of Manufacturing Employment Between Core
and Periphery Sectors for Metropolitan and Nonmetropolitan Areas, 1979.**

	INDUSTRY SECTOR		
LOCATION AND TYPE OF COUNTY[a]	Core[b]	Periphery[c]	Total
Metropolitan Statistical Area			
Large (Over 1 million pop.)			
Central City	79	21	100
Fringe	85	15	100
Medium (250,000 to 1 million pop.)	75	25	100
Small (Under 250,000 pop.)	72	28	100
Nonmetropolitan Area			
Urbanized (Urban pop. over 20,000)			
Adjacent to an SMSA	65	35	100
Not adjacent	58	42	100
Less Urbanized (Urban pop. 2500 to 20,000)			
Adjacent to an SMSA	53	47	100
Not adjacent	48	56	100
Rural			
Adjacent to an SMSA	40	60	100
Not adjacent	33	67	100

[a] For a description of categories, see Ross, *et al.*, 1979.
[b] Includes, following Beck, *et al.* (1978): Paper and Allied Products; Printing, Publishing
and Allied Products; Chemicals and Allied Products; Petroleum Refining and Related
Industries; Rubber and Miscellaneous Plastic Products; Primary Metal Industries;
Fabricated Metal Products; Machinery, Except Electrical; Electrical Machinery; Transportation Equipment; Stone, Clay and Glass Products; and Instrument Manufacturing.
(Classes based on U.S. Census industrial classification scheme.)
[c] Includes, following Beck, *et al.* (1978): Food and Kindred Products; Tobacco Manufacturing; Leather and Leather Products; Lumber and Wood Products; Furniture and Fix-
tures; Textile Mill Products; Apparel and Related Products; and Miscellaneous Manufacturing.

Source: Bureau of Economic Analysis, U.S. Department of Commerce.

while more rural and more remote locations tend to specialize in periphery sector manufacturing.[6]

A similar argument may be applied to the location of operations in multi-locational production firms. Headquarters generally need greater access to specialized knowledge and information than branch production operations and are therefore, as Lincoln (1978) has shown, more likely to be located in larger, more central places. Where urban headquartered firms have rural operations, they are likely to be periphery-type operations (although mining is an exception).

Being highly unionized, core-type employers are largely unaffected by any impacts that local economic growth may have on area wage rates. Thus the tension between exporting employers and central place business owners that characterizes rural area relations is less marked in urban locations. In fact, core operations are likely to benefit from local growth as it may increase access to information and specialized services. Core employers may be part of the local growth machine.

Rural-urban differences in the local interests of workers would seem to be less marked. Workers who belong to the strong nationally organized unions which characterize the core sector have, compared to less skilled workers, relatively weak local economic interests. Their incomes are determined less by local economic conditions than by conditions in the industry in which they work. However, while urban workers are more likely to work in a core-type setting than rural workers, many urban workers are in nonunionized, service sector jobs. It would not be correct, therefore, to characterize urban workers in general as having little to gain from local economic growth.

The second type of difference between contemporary rural and urban societies lies in their ecological organization, especially as it relates to the community of sentiment. As center size increases, areas of relatively intense social interaction do not expand commensurately. A large urban place will generally have many neighborhoods which serve as communities of sentiment for their residents. The spatial pattern of urban residential activities resembles the rural pattern in that the community of sentiment is much narrower geographically than the area in which people travel for goods and services (Hunter, 1975). However, the (larger) rural village is a focus of central place business interests as well as a focus of the community of sentiment. While there may be some urban neighborhood business interests, the main focus of urban central place business interests is the city itself. There is thus less blending of business interests with the community of sentiment in urban than in rural areas. An urban neighborhood "leader" is probably less likely to have economic interests attached to the locality than is a village "leader".

In addition, because of residential segregation by socioeconomic status, the community of sentiment associated with the urban neighborhood is likely to be more homogeneous with respect to the economic positions and interests of its residents than is the rural community of sentiment. Although mining and heavily industrialized rural areas may yield exceptions, there does not appear to be a rural counterpart to the urban "working class" neighborhood, at least at present. The rural and urban social milieus are thus quite different. The workers' residential segregation from business owners in urban areas suggests that the sense of identity as an economic interest group is likely to be stronger for urban than for rural workers, even for urban workers who are not union members.

The dearth of highly unionized industries, the common geographic focus of business interests with the community of sentiment, and the integration of economic groups in the community of sentiment may all help to make economic class less visible in rural than in urban areas, even though in terms of income distribution at least, inequality is greater in rural than in urban places (Duncan and Reiss, 1956; see McGranahan, 1980a). In sum, even though we have touched on only some aspects here, rural and urban communities continue to differ in important ways. The basis of these differences have, however, changed over time.

Conclusion

Rural sociology, particularly in the area of community research, tends to operate without the concept of economic class. Where local differences in role, behavior and attitude are anticipated, it is often on the basis of social background, length of residence, age, sex and status. These characteristics are quite relevant where residential activities and interests are concerned. Residential interests need not be directly related to economic class.

In addition to local interests associated with residence, however, are economic interests, which originate in the spatial segmentation of markets and vary with economic position. These interests have been central to the structure and dynamics of rural communities. They formed the basis of the conflict between town and country in the early agricultural communities. The central place interests of bankers, merchants and others with local trade areas have long been at the core of local efforts to attract outside industry. In an extension of plant-level interests, employers with extra-local sales have often played paternalistic roles in their local settings, and, with local labor market interests, these employers may act to constrain local economic growth. Of equal sociological importance are the growing importance of highly educated workers, whose local interests are almost exclusively residential,and the continuing lack of local labor market area representation of the interests of the less skilled workers. In short, the rural community and its changing structure cannot be understood without reference to class and economic interests.

Notes

1. This follows Weber's definition of class situation. For Weber, class exists when, "(1) a number of people have in common a specific causal component of their life chances insofar as (2) this component is represented exclusively by economic interests in the possession of goods and opportunities for income and (3) is represented under the conditions of commodity and labor markets" (1946:186). This paper is concerned with local class situations generated by local markets. Because I refer frequently to "interests", and the term, "class interests" is closely associated with Marxist theory, I use the terms "economic" and "economic interests" to avoid confusion.
2. Villages were often set off politically from the surrounding farm areas through their incorporation as independent municipalities.
3. Unfortunately, rural sociology post-dates the period of greatest conflict between town and country. This study was done in 1923-1925, at a time when local farmer-merchant antagonisms were beginning to abate and, as the authors note, transportation improvements were making towns less isolated from their hinterlands and from each other. Nevertheless, problems in village-country relations take up a chapter in the study.
4. It is not clear if this growth in manufacturing will continue during the present decade.
5. For recent work using the dual economy model, see Bloomquist and Summers (1982) and Beck, et al. (1978).
6. Table 9.1 is based on highly aggregated industrial categories, and the allocation to the core or periphery sector is in some cases arbritrary. Sharper geographic differences would presumably be found using more refined industrial categories. For a highly developed discussion of plant location, based on a somewhat different industrial typology, see Thompson (1965).

References

Baron, James N., and William T. Bielby. 1980. "Bring the firms back in: stratification, segmentation and the organization of work." *American Sociological Review* 45 (October): 737-765.

Beck, E.M.; Patrick M. Horan; and Charles M. Tolbert II. 1978. "Stratification in a dual economy: a sectoral model of earnings determination." *American Sociological Review* 43(October):704-720.

Bloomquist, Leonard E., and Gene F. Summers. 1982. "Organization of production and community income distributions." *American Sociological Review* 47(June): 325-338.

Brunner, Edmund deS., Gwendolyn S. Hughes, and Marjorie Patten. 1927. *American Agricultural Villages*. New York: George H. Doran Co.

———. and J.H. Kolb. 1933. *Rural Social Trends*. New York: McGraw-Hill.

———. and Irving Lorge. 1937. *Rural Trends in Depression Years: A Survey of Village-Centered Agricultural Communities 1930-1936*. New York: Columbia University.

Caudill, Harry M. 1963. *Night Comes to the Cumberlands.* Boston: Little, Brown and Co.

Chiswick, Barry R. 1974. *Income Inequality: Regional Analysis Within a Human Capital Framework.* New York: Columbia University Press.

Christaller, Walter. 1966 (1933). *The Central Places of Southern Germany.* Englewood Cliffs, NJ: Prentice-Hall.

Clemente, Frank; Dean Rojek; and E.M. Beck. 1974. "Trade patterns and community identity: five years later." *Rural Sociology* 39(Spring):92-95.

Clements, Donald W. 1978. "Utility of linear models in retail geography." *Economic Geography* 54(January):17-25.

Duncan, Otis Dudley, and Albert J. Reiss, Jr. 1956. *Social Characteristics of Urban and Rural Communities, 1950.* New York: John Wiley and Sons.

French, Robert M. 1969. "Change comes to Cornucopia: industry and community." Pp. 392-407 in R.M. French (ed.), *The Community: A Comparative Perspective.* Itasca, IL: F.E. Peacock.

Haga, William J., and Clinton L. Folse. 1971. "Trade patterns and community identity." *Rural Sociology* 36(March):42-51.

Haren, Claude C., and Ronald W. Holling. 1979. "Industrial development in nonmetropolitan America: a locational perspective." Pp. 13-46 in R.E. Lonsdale and H.L. Seyler (eds.), *Nonmetropolitan Industrialization.* Washington, DC: V.H. Winston & Sons.

Hawley, Amos H. 1941. "An ecological study of human service organization." *American Sociological Review* 6(October):629-639.

Hunter, Albert. 1975. "The loss of community: an empirical test through replication." *American Sociological Review* 40(October):537-552.

Johansen, Harley E., and Glenn V. Fuguitt. 1973. "Changing retail activity in Wisconsin villages 1939-1954-1970." *Rural Sociology* 38(Summer):207-218.

Johnson, Kenneth M. 1982. "Organizational adjustment to population change in nonmetropolitan America: a longitudinal analysis of retail trade." *Social Forces* 60:1123-1139.

Kolb, J.H. 1957. "Neighborhood-family relations in rural society: a review of trends in Dane County, Wisconsin over a 35 year period." *Research Bulletin* 201, University of Wisconsin-Madison Agricultural Experiment Station.

_____. and LeRoy J. Day. 1950. "Interdependence in town and country relations in rural society." *Research Bulletin* 172. University of Wisconsin-Madison Agricultural Experiment Station.

Lincoln, James. 1978. "The urban distribution of headquarters and branch plants in manufacturing: mechanisms of metropolitan dominance." *Demography* 15(May):213-222.

Lipset, Seymour Martin. 1968. *Agrarian Socialism: The Cooperative Commonwealth Federation in Saskatchewan: A Study in Political Sociology.* Garden City, NY: Doubleday and Co. (Anchor Books).

McGranahan, David A. 1980a. "The spatial structure of income distribution in rural regions." *American Sociological Review* 45(April):313-324.

_____. 1980b. "Changing central place activities in northwestern Wisconsin." *Rural Sociology* 45(Spring):91-109.

_____. 1982. "Class, space and community structure." Unpublished manuscript.

Molotch, Harvey. 1976. "The city as a growth machine: toward a political economy of place." *American Journal of Sociology* 82(September):309–332.

Newby, Howard. 1980. "Rural sociology—a trend report." *Current Sociology* 28(Spring): 1–141.

————, and Frederick H. Buttel. 1980. "Toward a critical rural sociology." Pp. 1–35 in F.H. Buttel and H. Newby (eds.), *The Rural Sociology of the Advanced Societies: Critical Perspectives.* Montclair, NJ: Allanheld, Osmun & Co. Publishers.

————; Peter Saunders; David Rose; and Colin Bell. 1979. "In search of community power." Pp. 121–144 in G.F. Summers and A. Selvik (eds.), *Nonmetropolitan Industrial Growth and Community Change.* Lexington, MA: D.C. Health and Co.

Ross, Peggy J.; Herman Bluestone; and Fred Hines. 1979. "Indicators of social well-being for U.S. counties." *Rural Development Research Report* No. 10, Economic Development Division, Economic Research Service, U.S. Department of Agriculture.

Sanderson, Dwight. 1939. "Locating the rural community." *Cornell Extension Bulletin* 413, Ithaca: NY State College of Agriculture at Cornell University. Excerpted (pp. 164–168) in R.L. Warren (ed.), *Perspectives on the American Community,* Second Edition (1973). Chicago: Rand McNally College Publishing.

Smith, Carol A. 1972. "Exchange systems and the spatial distribution of elites: the organization of stratification in agrarian societies." Pp. 309–374 in C.A. Smith (ed.), *Regional Analysis: Volume II, Social Systems.* New York: Academic Press.

Suttles, Gerald, and Morris Janowitz. 1979. "Metropolitan growth and democrative participation." Pp. 157–178 in A.H. Hawley (ed.), *Societal Growth: Processes and Implications.* New York: The Free Press.

Summers, Gene F.; Sharon D. Evans; Frank Clemente; E.M. Beck; and Jon Minkoff. 1976. *Industrial Invasion of Nonmetropolitan America: A Quarter Century of Experience.* New York: Praeger Publishers.

Thompson, Wilbur R. 1965. *A Preface to Urban Economics.* Baltimore: John Hopkins University Press.

U.S. Bureau of Census. 1972. Census of Population: 1970 General Social and Economic Characteristics. Final Report PC (1) C1 United States Summary. Washington, DC: U.S. Government Printing Office.

Warner, W. Lloyd, and J.O. Low. 1946. "The factory and the community." Pp. 21–45 in W.F. Whyte (ed.), *Industry and Society.* New York: McGraw-Hill.

Warren, Roland. 1978. *The Community in America.* Third Edition. Chicago: Rand McNally College Publishing.

Weber, Max. 1946. "Class, status and party." Pp. 180–195 in H. Gerth and C.W. Mills, *From Max Weber: Essays in Sociology.* New York: Oxford.

Weist, Edward. 1923. *Agricultural Organization in the United States.* Lexington, KY: Kentucky University Press.

Wilkinson, Kenneth P. 1979. "Rural community change." Pp. 115–125 in T.R. Ford (ed.), *Rural U.S.A.: Persistence and Change.* Ames, IA: Iowa State University Press.

Chapter 10

Farm Family and the Role of Women

Wava Gillespie Haney

The conjunction of women, work, and family is a research area with a limited history. The interdependence of the work world and the family has been virtually ignored by researchers (Kanter, 1977). Research on work has focused primarily on organizations and the men within them while research on family has concentrated mostly on relationships between women and children. Furthermore, these research traditions have reinforced a methodological focus on the individual as the unit of analysis and have fostered an assumpton that in the industrial world work and family were separate domains – of men and women, respectively. Kanter summarized this "myth" succinctly: "In a modern industrial society work life and family life constitute two separate and non-overlapping worlds, with their functions, territories and behavioral rules" (1977:8). The point of Kanter's critique is academic: we cannot hope to understand either the economy or the family and their dimensions if we fail to take into account the entire cast of actors and the interrelationships between institutions. [1] Obviously, Kanter's concerns are relevant to studies of any sector of the economy, but they are particularly applicable to the agricultural sector where family-centered production units predominate. Ironically, studies in the agricultural sector have historically raised few questions about the interdependence of production and family systems and of specific activities of women and men. [2] Work-family relationships in farm family studies, for example, were generally limited to the impact of farm wives' attitudes and helpmate activities on farm production. In short, women were treated as "factors" in male producers' success (Joyce and Leadley, 1977).

The author thanks Cornelia Butler Flora, Emil Haney, Gene Summers, and two anonymous reviewers for helpful comments on an earlier draft.

Technology and associated changes in the production relationships in agriculture, in particular, and the economy, in general, not only have important consequences for women who share work and/or family space with other members of the production unit, but such changes are also importantly affected by the activities of women. In order to understand an agricultural economy and the economic system in which it is embedded, it is necessary to look at the market and nonmarket activities of both women and men. And in order to understand what is happening to farm women, one must examine the changing structure of agriculture. This paper will examine theoretically this mutually dependent relationship in the context of an agricultural system undergoing significant technological and social change, including expanded scale of production; increased separation of control over the factors of production; increased levels of capital investment and debt load; greater integration of production, processing, and marketing functions; increased substitution of capital for human labor and wage labor for family labor; and increased part-time farming (cf., Rodefeld, 1978; Newby, 1980; Wilkening, 1981; Havens, 1982).

In the past decade of scholarship on women comparative studies of history and society suggest that technological and related developments associated with the emergence of industrial capitalism tended to marginalize women by reducing women's role in the production of socially necessary goods and services (Boserup, 1970; Sachs, 1975), women's control over productive technology (de Beauvoir, 1952), women's control over the distribution of social goods (Friedl, 1975), and the mutual economic dependency of men and women (Sanday, 1975). A change in the degree of women's contribution to societal production acccompanied the increased use of labor-saving technology (Bossen, 1975) and the development of state societies (Reiter, 1975; Rosaldo and Lamphere, 1975).

Historical evidence from the United States suggests that women played a vital role in production under conditions of labor scarcity and limited use of technology—for example, colonial and frontier America (Ryan, 1975). In industrial America, however, the decline in family-centered production, the continual division and sub-division of labor, and the application of greater quantities of capital-intensive technology curtailed women's productive role in society. Married women were confined to production for household use while their consumer role increased (Tilly and Scott, 1978; Weinbaum and Bridges, 1979). Single women were employed almost exclusively in clerical work, retail sales, teaching, and nursing (Ryan, 1975; Hartmann, 1976). To be sure, there were periods of great labor demand, such as during major wars, which lessened the grip over this ghettoization of women's labor. During World War II, for example, women entered traditional male jobs in large numbers (Friedan, 1963). And since the late 1950s women, including married women with young children, have entered the paid labor force in vast numbers (Almquist, 1977). Today, women are concentrated in some 20 of 250 occupations listed by the Census Bureau (Blau, 1979). Whatever the nature of women's production activities, however, they have remained the guardians of the reproduction of social institutions (Ryan, 1975; Smith, 1975-76).

Rural women in the United States, including those who live and work on farms, have followed this national trend of increased paid labor force participation (Sweet, 1972; Bokemeier et al., 1980), often while concurrently contributing to a wide range of production activities on the farm (Dorner and Marquardt, 1977; Wilkening and Ahrens, 1979; Fassinger and Schwarzweller, 1980; Bokemeier and Coughenour, 1980; Kada, 1980; Jones and Rosenfeld, 1981; Maret and Copp, 1982). Significant numbers of farm women integrate productive and reproductive responsibilities in work weeks of up to 110 hours (Boulding, 1979) and with involvement in community and farm organizations and agencies (Jones and Rosenfeld, 1981). A growing number of women – single, married, and widowed – are sole or principal farm operators or managers (Pearson, 1979; Kalbacher, 1982). Women's involvement in agricultural production often begins early – for example, the contributions of daughters (Colman et al., 1975a, 1975b, 1976, 1978a, 1978b; Wilkening and Ahrens, 1979) – and continues into retirement – for example, the role of widows in intergenerational farm transfers and farm management (Salamon and Keim, 1979). Finally, women form an important part of the migrant labor force (Hacker, 1980) and the labor force of agribusiness firms (Friedland et al., 1980). As individuals, then, farm women are operators, managers, laborers, landlords, and partners in widely varied family-centered agricultural production units. The most numerous group of farm women are partners in family production units.

While recent literature has expanded our general knowledge of farm women, it seems reasonable to conclude with Joyce and Leadley (1977), Hill (1981), and others that farm women, even farm wives, remain virtually invisible in American social science. Yet, it is clear that women are principal actors in the productive[3] and reproductive aspects of the farming system and, therefore, must be an integral part of an analysis of the food system and its production units. Space does not permit consideration of the entire range of women's involvement in United States agricultural production and an historical perspective. This paper focuses on the role of women in the nearly three-fourths of the U.S. agricultural production units classified as marginal in terms of total product output (USDA, 1981). Effectively, this excludes from consideration farm women who only provide large amounts of land or capital to the agricultural production process and those who work exclusively as agricultural wage laborers or salaried managers.

Women and the Structure of Agriculture

The profile of American agriculture presented in the USDA report, *A Time to Choose,* divides America's farms into four categories. Farms with sales of $40,000 or more per year (tabbed "full time" and "super farms" by USDA) constitute about 22 percent of the total number of farms and produce approximately 80 percent of the farm output. At the other extreme, 44 percent of the farms have sales under $5,000 per year and produce only 2 percent of the total farm output ("rural farm residences"). An intermediate category ("small farms") accounts for 34 percent of the farms which produce 17 percent of the total output. "If the report can be said to have a theme, it is that the periphery of American agriculture is growing," writes Strange (1981:18) in a recent review. Clearly, there has been a continuing decline in the number of family-centered production units whose major source of income for and employment of family members is the farm. Even more striking is the decline in the importance of "family farms" as the source of the nation's food supply. Yet, in some commodities and in certain regions small farms persist and are increasing in number (Harper et al., 1980) though their characteristics and those of their proprietors are quite diverse (Heffernan et al., 1982).

Another way to view U.S. agriculture is to examine landholding patterns and the ownership and control of other inputs (Rodefeld, 1978; Goss et al., 1980; Moore and Wilkening, 1981). In his analysis of occupational positions in the contemporary United States, Wright (1979) isolated three dimensions of control over productive resource allocation in addition to ownership—control over what is produced, control over how factors of production are used, and control over the labor process. With these dimensions in mind, Wright classified occupations into three general class positions—capitalist, proletariat, and petty bourgeoisie—and three contradictory class locations—small employer, semi-autonomous employee, and manager. He assigns agricultural producers to the petty bourgeoisie.

But the production unit and occupational literature converge to suggest that agricultural producers in the United States occupy diverse class positions. For example, "factory farms" (Rodefeld, 1978) and "primary farms II" (USDA, 1981) which hire and control a large labor force to operate and manage them contain two major social classes. Likewise, "larger than family farms" (Rodefeld, 1978) incorporate small employers and permanent wages laborers. Using Wright's criteria, however, the large and apparently growing small farm sector of U.S. agriculture is neither homogeneous in its production and occupational characteristics nor petty bourgeois. An analysis using Wright's four dimensions of resource control separates the position of ownership of land and other fixed inputs, control over production and investment decisions, and the use of one's own and usually one's family's labor—the independent family farm—from the position of partial control over what and how to produce and/or the physical means to produce it with one's own and usually one's family's labor—the semi-autonomous position of tenants, renters, and contract farmers. The independent family farmer, or in Wright's terminology, the petty bourgeoisie of agriculture, is virtually nonexistent. On the contrary, renters, contract farmers and other forms of family autonomous relations are growing in U.S. agriculture and may occupy, as Strange contends, the production periphery. The characteristics Wright attributes to semi-autonomous employee production, however, are not only important to the analysis of farm families' relationship to agricultural production[4], but also to the analysis of the overall production relations of the slightly over three-fourths of the farms where family-centered production is often combined with off-farm employment.

As individuals and/or families, farm people relate to the processes of production in multiple and often contradictory ways. Farm women who are partners in family-centered production units work both off and on the farm or may be employed solely off the farm. Women involved in both may relate to the agricultural production in one way and to non-agricultural production in yet another way. The same is true of farm men, including farm husbands. If there is a trend in the limited empirical data, it is that multiple and contradictory relations are increasing within farm families. Thus, to understand changes in the overall productive structure of agriculture and to understand the work situation of farm women (and farm men), we must consider the multiple dimensions of control over production in the various productive roles of each member of the household.

To analyze the role of women in the farm family, we must add the dimension of reproduction because women's work and women's roles within the family are certainly not limited to farm and off-farm work. There is overwhelming agreement in the literature that women are still almost solely responsible for household and family maintenance, with peak demands under conditions of adversity (Weinbaum and Bridges, 1979), as well as for child rearing, family and community relations, and other reproductive functions (Smith, 1975-76).

In sum, adequate consideration of the role of women within the farm and the family and adequate analysis of the food production system and its production units necessitates a conceptual framework which sees work and family as interpenetrating and mutually dependent within the context of a household analysis. Within this framework, it is important to delineate major work and family patterns in response to and in their impact upon the changing conditions of the food and general economic system. This paper focuses on these general issues.

Women, Work, and Family on the Agricultural Periphery

According to the USDA (1981), slightly more than three-fourths of the agricultural production units are marginal in terms of their contribution to total farm output. While this periphery of American agriculture no doubt includes some family farms of yore, increasingly the farms of the periphery can be divided into two kinds of units: those operated by families who supplement agricultural work and income with off-farm employment and off-farm employees who supplement their work and income with farm production.[5] For the former, at least, off-farm employment seems to represent an adaptation in the organization of the household production unit to the deteriorating terms of trade for agricultural producers.[6]

Most of these family-centered production units are dwarfed by a giant service apparatus which supplies them inputs and buys their products. Concomitantly, many of these farms are pressed by the suburbanization process: demand for farmland by housing and commercial developments and highway construction has inflated land prices, raised property taxes, and created a growing scarcity of readily available and reasonably priced fertile land. At the same time, modern production practices deplete soil and water resources and threaten productivity levels, lest there be ever-increasing levels of purchased yield-increasing inputs. Moreover, agricultural producers have become increasingly integrated into world markets and dependent upon American foreign policy, especially during the past decade in commodities such as grain. In sum, widely fluctuating product prices, triggered in part by shifts in foreign policy, coupled with soaring input prices have put a severe squeeze on net farm income. Barring any major reductions in output from unfavorable weather conditions, the "free market" emphasis of the current administration could very well exacerbate short-term farm surpluses and income levels as farm families seek to offset reduced profit margins with increases in output.

In general, these conditions have made and will continue to make farm families more dependent upon the work of farm women — wives, daughters, and daughters-in-law — in farm and household production and outside employment.[7] Like many nonfarm families, an increasing number of farm families have abandoned the ideal in our society since the early nineteenth century — the male as producer, the female as housewife and consumer. Trends within agriculture together with movement to the countryside of industries which employ a large female labor force and the rapid expansion of traditional female jobs in public and private services are also instrumental in farm women's entering production roles outside agriculture. It is likely that many men will continue to farm because their wives have off-farm employment. Other families will stay in farming because farm women assume major production responsibilities while their husbands engage in nonfarm employment. Given the role of education as an occupational gatekeeper, it seems clear that the wife's education is a central factor in determining both the type of income-generating activities combined with agriculture and the distribution of work roles within and outside the family. Household composition also affects the supply of family labor and thus its allocation between and within income-generating activities.

In many of the family production units where ownership of land and other capital assets remains with a landlord or encumbered to financial institutions and/or where returns from corporate buyers and contractors are low, farm women will not only increase their labor and management input to the farm enterprise, but will also be pressed to increase their production for household use as well as off-farm employment, where available.[8] Or, farm women may become more centrally involved in production by providing more labor to the farm even at the expense of family and community functions. For example, women's labor may be substituted for hired male labor while the family relies more on purchased services and the community's volunteer services are curtailed. Or high capital costs like currently exist undoubtedly will mean that farm families alter their forms of production to substitute capital for women's labor and management skills. At the same time, the range of women's functions in the production unit, as with those of men, will begin to narrow.

In other words, farm women's labor is vital to the survival strategies of farm families and family farms. These strategies, in turn, are important to understanding changing production and class structures in agriculture. We look next at women's role in family survival strategies.

Women and the Agricultural Production Unit

Concentration and centralization of the food production system have eroded the degree of control over productive factors by family-centered production units and thus limited the range of positions available in the organization of these production units. One set of factors which importantly influences the work of farm women, and which in turn are influenced by farm women's individual characteristics, is the choices made at the production unit level. Of particular significance is: (1) the allocation of the sources between market production and production for home use, (2) the choice of principal commodities and the technologies used to produce them, and (3) the level of the family inventory of productive stocks and their access to other inputs. The choices of the production unit are also circumscribed by its terms of trade in the sector and product market which, in turn, are substantially influenced by state policies.

Likewise, the existence, nature, and amount of off-farm employment and the relative opportunity costs for various family members are of primary importance to a family's particular survival strategy and thus to the role of farm women. Finally, family composition and individual characteristics of women will have an impact upon the internal organization of farm family businesses; these include the age and sex composition of children, the existence of intergenerational partnerships, and the educational level and work history of women. This section identifies some of the type of relationships extant among these sectors and the way in which they may circumscribe farm women's market and non-market work.

Orientation to production distinguishes those for whom agricultural enterprises are viewed as businesses from those for whom the primary focus of production is a lifestyle or for growing home food and for whom sales of surplus are secondary or perhaps planned in order to qualify for tax benefits. Lifestyle-oriented production is likely to encompass a disproportionate number of families whose major source of income is off-farm employment. Many of these families depend upon employment in industrial and service jobs, but an increasing proportion seems to be college-trained professionals. On farms oriented to production for use or enjoyment, wives and daughters with higher educational levels are (1) likely to have nonfarm employment, and (2) likely to be involved in farm work to the degree of their interest in farming and in the principal commodities produced. Still, their contribution to production is likely to be substantial since this type of production is often labor intensive. Older women with lower levels of education are more likely to be substantially involved in food production and less likely to have off-farm employment.

On the other hand, in production units that are totally or principally market oriented one is likely to find farm work to be important in absorbing family labor and agricultural income to constitute a large proportion of total income. Moreover, the nature of the commodities produced will be important in the distribution of women's off-farm and farm work since the nature of the principal product influences the absorptiveness and timing (Kanter, 1977) of agricultural work and thus circumscribes family members' options. The combination of livestock production, such as dairying, with seasonal cash crops creates peak periods of very absorptive work but a steadiness even in the "off season." This combination of products is likely to mean seasonal adjustments in women's work in the enterprise and probably an attempt to create corresponding rhythms in any nonfarm employment.

In market-oriented production units with variable yet constant demands for labor, women would more likely supply labor on a permanent basis as well as in peak periods. Moreover, these women probably supply management skills throughout the year. In contrast, seasonal cash cropping, which is common in the corn belt and wheat growing areas, would seem to lend itself to a farm/nonfarm rhythm to correspond with the cropping season. Given the level and kind of mechanization of these types of cash crops, the degree of women's involvement in production activities may be low even during seasonal peaks, but her farm management and family maintenance work will be high.

Educational level of the wife is central to the allocation of the principal off-farm employment role to the husband or the wife and to her type of off-farm employment (Wilkening and Ahrens, 1979; Moore and Wilkening, 1981). With increased labor force participation of mothers, even those with young children, educational levels of the wife are central throughout the family life cycle to the distribution of work roles within and outside the family. When women have high educational levels, they are more likely to work off the farm and more likely to work in services or the professions. Although women usually have more formal education and thus a better opportunity to find employment, men can usually find jobs that pay more than women. Apparently, occupational and wage discrimination by sex is important in husbands working off the farm more often than wives (Jones and Rosenfeld, 1981).

Off-farm employment of wives with higher educational levels, however, does not necessarily lead to limited involvement in the farm enterprise. On the contrary, the available data suggest that such farm women are likely to fulfill a variety of functions in the agricultural firm. In fact, it is not uncommon for such women to provide an important source of working and investment capital for the production unit, to be involved in at least some management, and at the same time provide occasional labor to the enterprise and almost all the labor for the household. Not surprisingly, Linn (1982) found that husbands were more satisfied with women's work on the farm than women were with husbands' work in the household.

It has been generally assumed that stage of the family life cycle has an important impact upon the amount and kind of contribution wives make to the internal division of labor in the agricultural enterprise. As children, both sons and daughters, grow older, they are likely to take on more farm tasks. What remains unclear are the conditions under which a labor contribution by children substitutes or complements that of their parents. Indeed, minimal change in the amount of labor or the functions performed by mothers may result from the expansion of the enterprise as the family labor supply grows. For example, the literature suggests there is no change in farm wives' involvement in management functions such as record keeping (Wilkening and Ahrens, 1979) nor in the amount or kind of labor supplied to the farm by mothers (Fassinger and Schwarzweller, 1980) at different stages of the life cycle.

Greater labor force participation by mothers and the coincidence of family labor supply and farm expansion seem to weaken the interaction between life cycle and work allocation in the early and middle stages of the family cycle. In the later stages, however, women's roles in intergenerational farm transfers need to be examined. While roles will vary with development cycles of farm transfers, it is likely that during this incorporating period farm women's productive work diminishes while their reproductive work increases. More specifically, a wife's/mother's labor input to the agricultural enterprise may change from permanent laborer to occasional or seasonal worker, but this decrease may be offset by an expansion of her management function. Some have suggested (Kohl, 1976; Rosenblatt and Anderson, 1981) that the wife/mother serves as a fulcrum of communication between father and son within the family and carries that same function into nearly all phases of personnel and production management within the enterprise. The incorporating period can often mean that the family production unit is multi-household. In this type of unit there is a need to examine the work allocation with the farm enterprise of not only the wife/mother but the daughter(s) and/or daughter(s)-in-law. In such cases, functions may be greatly influenced by the legal arrangements for the production unit as a whole, by the educational levels of the women involved, and by the values of ethnic communities (Salamon and O'Reilly, 1979; Salamon and Keim, 1979).

Women and Agricultural Policy

Farm women's centrality, together with their social role in production, has generated greater self-esteem[9] (Jones and Rosenfeld, 1981) and a tendency to organize for change and to seek political power as farm women. Overall, however, farm women's involvement in and knowledge of farm programs and farm organizations lags behind farm men (Jones and Rosenfeld, 1981). Visibility of farm women in agricultural organizations and agency programs along with a surge in farm women's organizations show their involvement in the policy process. But our central concerns about the impact of farm women on agricultural structure and policies which circumscribe that structure are the nature of farm women's consciousness and the form of their involvement in the political arena. At the descriptive level, the major research issue is the focus of women's political involvement – e.g., along sex, class, interest group lines, or in combination (cf. Hill, 1981; Flora, 1981; Elbert, 1981). Analytically, our concern is how structural and individual characteristics frame political consciousness and action of farm women.

Domestic and foreign policies and general economic conditions which threaten the survival and reproduction of family-centered production units trigger political action by farm family members including women. For individual farm women degree of identification with the agricultural production unit and type of functions they perform in agricultural production and in the farm family are important in shaping both their level of political involvement and the type of policies they advocate.

Farm women who strongly identify with and actively manage and operate family-centered production units and who are committed ideologically to the family farm are more likely to frame their activism in class and interest group terms (Flora, 1981). They are likely to participate in general farm organizations and commodity groups which advocate policies intended to preserve farming as a family business (e.g., repeal of a grain embargo; restoration of price supports or marketing orders; lower interest rates and property taxes). Farm women's organizations often combine political action along class and interest group lines with concerns about their rights as women involved in family enterprises (e.g., inheritance and tax laws, discriminatory policies among agricultural service agencies and private businesses serving farm families).

An emerging issue, however, is how increasing dependence upon land, capital, technology, and management from outside the family will change farm families' and farm women's policy orientations. With increases in part-time farming combined with off-farm employment, the degree of complementarity between the location of farm and off-farm work in the overall occupational structure will be important in shaping political action of farm women and men.

Conclusion

Changes in the structure of agriculture have necessitated farm women's increased involvement in production. Meanwhile, the tendencies toward a substitution of public services for farm women's traditional family and community activities and the involvement of farm men in household maintenance and child rearing appear to be slow. Women continue to play a central role in farm families' and rural communities' survival and reproduction strategies. This means that agricultural social scientists cannot possibly understand either the farm family or the production unit without taking into account the activities of farm women. At the macro level, the process of social differentiation within U.S. agriculture and the changing class structure of rural areas can only be analyzed through a consideration of farm families' total productive relation.

The changing structure of agriculture that created the necessity for farm women to enter income generating activities on or off the farm occurred in the context of an occupational structure highly segregated by sex and an agricultural service structure almost exclusively delivered by males to males. Thus, scholars of contemporary political and social movements in U.S. agriculture must give more attention to the interaction of class, interest group, and sex-based orientation among farm women.

In short, this paper is much more than a call for a sociology of rural women. Rather, it is an assertion that the sociology of agriculture can be advanced only through a more complete and systematic understanding of the role of farm women in agricultural production and policy making.

Notes

1. This is in contrast to a feminist position that women's work should be acknowledged and appreciated. This distinction is an important one since writing about women by women is often interpreted to be simply feminist.
2. The research of Eugene A. Wilkening and his associates on farm women and farm families stood virtually alone until the 1970s. See Wilkening, 1958; Wilkening and Bhradwaj, 1967, 1968; Wilkening and Guerero, 1969.
3. In their national study Jones and Rosenfeld found that "very few of the women in our sample are totally uninvolved in the work and management of their operations" (1981:52).
4. During the revision of this paper, I discovered that Patrick Mooney and I had independently applied Wright's analysis to U.S. agriculture. For Mooney's application, see "Class Relations and Class Structure in the Midwest," mimeographed (Madison:

Department of Rural Sociology, University of Wisconsin). This is an earlier draft of "Toward a Class Analysis of Midwestern Agriculture," presented at the Rural Sociological Society meetings in San Francisco, September 1982.

5. My division seems consistent with what Heffernan and his associates found in their Missouri study. They write: "There was a very strong tendency for small-farm families to receive either little or most of their family income from nonfarm sources. . ." (Heffernan, et al., 1982:66).

6. Non-agricultural employees entering farming may also be adapting to these deteriorating terms of trade, but form the position of having anticipated difficulty in generating an adequate income from farming and therefore seeking more remunerative and/or predictable income-generating occupations.

7. In a currrent field study of Wisconsin farm families, Wilkening (personal communication) found that during the present economic downturn men are more likely to be laid off than women. Thus women are providing the bulk of the family's off-farm income.

8. Wilkening (personal communication) found a positive relationship between women working off the farm and the amount of farm indebtedness, but no relationship between women's off-farm employment and nonfarm indebtedness.

9. Jones and Rosenfeld (1981) report 60 percent of the married women felt they could operate the farm without their husbands.

References

Almquist, Elizabeth M. 1977. "Review essay: women in the labor force." *Signs* 2 (Summer): 843-855.

Blau, Francine D. 1979. "Women in the labor force: an overview." Pp. 265-289 in Jo Freeman (ed.), *Women: A Feminist Perspective*. Palo Alto, CA: Mayfield.

Bokemeier, Janet, and C. Milton Coughenour. 1980. "Men and women in four types of farm families: work and attitudes." Paper presented at the Rural Sociological Society meetings, Ithaca, NY.

Bokemeier, Janet, Verna Keith, and Carolyn Sachs. 1980. "What ever happened to rural women?" Paper presented at the annual meetings of the Rural Sociological Society, Ithaca, NY.

Boserup, Ester. 1970. *Woman's Role in Economic Development*. London: George Allen and Unwin.

Bossen, Laurel. 1975. "Women in modernizing societies." *American Ethnologist* 2 (November): 587-601.

Boulding, Elise. 1979. "The labor of farm women in the United States: a knowledge gap." Paper presented at American Sociological Association meetings, Boston, MA.

Colman, P. Gould, and Jean Lowe. 1975a. "How farm families make decisions: the Brauns." *Rural Sociology Bulletin* 79, A.E. Res. 75-30 (October). Cornell University, Ithaca, NY.

Colman, P. Gould, and Sarah Elbert. 1975b. "How farm families make decisions: the Roots." *Rural Sociology Bulletin* 79, A.E. Res. 75-34 (December). Cornell University, Ithaca, NY.

Colman, P. Gould, Sarah Elbert, and Joyce Finch. 1976 ."How farm families make decisions: the Neirikers." *Rural Sociology Bulletin* 79, A.E. Res. 76-11 (June). Cornell University, Ithaca, NY.

Colman, P. Gould, Laurie Konigsburg, and Leslie Puryear. 1978a. "How farm families make decisions: the Nordahls." *Rural Sociology Bulletin* 79, A.E. Res. 78-5 (May). Cornell University, Ithaca, NY.

Colman, P. Gould and Leslie Puryear. 1978b. "How farm families make decisions: the Sawyers." *Rural Sociology Bulletin* 79, A.E. Res. 78-3 (January). Cornell University, Ithaca, NY.

de Beauvoir, Simone. 1952. *The Second Sex.* New York: Bantam.

Dorner, Peter, and Mark Marquardt. 1977. "Economic changes on a sample of Wisconsin farms: 1950-1975." Agricultural Economics Staff Paper no. 135. Department of Agricultural Economics, University of Wisconsin-Madison.

Elbert, Sarah. 1981. "The challenge of research on farm women." *Rural Sociologist* 1 (November): 387-390.

Fassinger, Polly A., and Harry K. Schwarzweller. 1980. "Exploring women's work roles on family farms: a Michigan case study." Paper presented at Rural Sociological Society meetings, Ithaca, NY.

Flora, Cornelia Butler. 1981. "Farm women, farming systems, and agricultural structure: suggestions for scholarship." *Rural Sociologist* 1 (November): 383-386.

Friedan, Betty. 1963 *The Feminine Mystique.* New York: Dell Publishing.

Friedl, Ernestine. 1975. *Women and Men: An Anthropologist's View.* New York: Holt, Rinehart and Winston.

Friedland, William H., Mena Furnari, and Enrique Pugliese. 1980. "The labor process in agriculture." Paper presented at Working Conference on the Labor Process, Santa Cruz, CA.

Goss, Kevin F., Richard D. Rodefeld, and Frederick H. Buttel. 1980. "The political economy of class structure in U.S. agriculture: a theoretical outline." Pp. 82-132 in Frederick H. Buttel and Howard Newby (eds.), *The Rural Sociology of the Advanced Societies.* Montclair, N.J.: Allanheld, Osmun.

Hacker, Sally. 1980. "Technological change and women's role in agribusiness." *Human Services in the Rural Environment* 5 (January-February):6-14.

Harper, Emily B., Frederick C. Fliegel, and J. C. van Es. 1980. "Growing numbers of small farms in the North Central States." *Rural Sociology* 45 (Winter): 608-620.

Hartmann, Heidi. 1976. "Capitalism, patriarchy, and job segregation by sex." *Signs* 1 (Spring): 137-169.

Havens, A. Eugene. 1982. "The changing structure of U.S. agriculture." Pp. 308-316 in Don A. Dillman and Daryl J. Hobbs (eds.), *Rural Society in the U.S.: Issues for the 1980s.* Boulder, CO: Westview Press.

Heffernan, William D., Gary Green, Paul Lasley, and Michael F. Nolan. 1982. "Small farms: a heterogeneous category." *Rural Sociologist* 2 (March): 62-71.

Hill, Frances. 1981. "Farm women: challenge to scholarship." *Rural Sociologist* 1 (November): 370-382.

Jones, Calvin, and Rachel A. Rosenfeld. 1981. "American farm women: findings from a national survey." *NORC Report* no. 130. Chicago: National Opinion Research Center.

Joyce, Lynda M., and Samuel M. Leadley. 1977. "An assessment of research needs of women in the rural United States: literature review and annotated bibliography." AE&RS 127. Mimeographed. University Park: Pennsylvania State University.

Kada, Ryohei. 1980. *Part-Time Family Farming: Off-Farm Employment and Farm Adjustments in the United States and Japan.* Tokyo: Center for Academic Publications.

Kalbacher, Judith Z. 1982. "Women farmers in America." USDA Economic Research Service, Washington, D.C.

Kanter, Rosabeth Moss. 1977. *Work and Family in the United States: A Critical Review and Agenda for Research and Policy.* New York: Russell Sage Foundation.

Kohl, S.B. 1976. *Working Together: Women and Family in Southwestern Saskatchewan.* Toronto: Holt, Rinehart and Winston of Canada.

Linn, Gary J. 1982. "Task performance and decision making as related to marital satisfaction in Wisconsin farm families." Paper presented at Rural Sociological Society meetings, San Francisco.

Maret, Elizabeth, and James H. Copp. 1982. "Some recent findings on the economic contribution of farm women." *Rural Sociologist* 2 (March): 112-115.

Moore, Keith H., and Eugene A. Wilkening. 1981. "Family farm typology for the analysis of family roles and attitudes." Paper presented at Rural Sociological Society meetings, Gwelph, Ontario, Canada.

Newby, Howard. 1980. "Rural sociology—a trend report." *Current Sociology* 28 (Spring): 1-141.

Pearson, Jessica. 1979. "Note on female farmers." *Rural Sociology* 44 (Spring): 189-200.

Reiter, Rayna R. 1975. "Men and women in the south of France: public and private domains." Pp. 252-282 in R.R. Reiter (ed.), *Toward an Anthropology of Women.* New York: Monthly Review Press.

Rodefeld, Richard D. 1978. "Trends in U.S. farm organizational structure and type." Pp. 158-177 in R.D. Rodefeld, et al. (eds.), *Change in Rural America: Causes, Consequences and Alternatives.* St. Louis: C.V. Mosby Co.

Rosaldo, Michelle Zimbalist and Louise Lamphere. (eds.) 1975. "Introduction." Pp. 1-15 in *Woman Culture and Society.* Stanford, CA: Stanford University Press.

Rosenblatt, Paul C. and Roxanne M. Anderson. 1981. "Interaction in farm families: tension and stress." Pp. 147-166 in Raymond T. Coward and William M. Smith (eds.), *The Family in Rural Society.* Boulder, CO: Westview Press.

Ryan, Mary P. 1975. *Womanhood in America: From Colonial Times to the Present.* New York: New Viewpoints.

Sachs, Karen. 1975. "Engels revised: women, the organization of production, and private property." Pp. 207-222 in M.Z. Rozaldo and L. Lamphere (eds.), *Woman Culture and Society.* Stanford, CA: Stanford University Press.

Salamon, Sonja, and Ann Mackey Keim. 1979. "Land ownership and women's power in a midwestern farming community." *Journal of Marriage and the Family* 41 (February): 109-119.

Salamon, Sonja and Shirley M. O'Reilly. 1979. "Family land and developmental cycles among Illinois farmers." *Rural Sociology* 44 (Fall): 525-542.

Sanday, Peggy R. 1975. "Female status in the public domain." Pp. 189-206 in M.Z. Rosaldo and L. Lamphere (eds.), *Woman Culture and Society.* Stanford, CA: Stanford University Press.

Smith, Dorothy. 1975-76. "Women, the family and corporate capitalism." *Berkeley Journal of Sociology* 20: 55-90.

Strange, Marty. 1981. "Radical questions, less than full answers in Bergland's report." *Food Monitor* 23 (July/August): 18-23.

Sweet, James A. 1972. "The employment of rural farm wives." *Rural Sociology* 37 (December): 553-577.

Tilly, Louise A. and Joan W. Scott, 1978. *Women, Work and Family.* New York: Holt, Rinehart and Winston.

U.S. Department of Agriculture (USDA). 1981. *A Time to Choose.* Washington, D.C.: U.S. Government Printing Office.

Weinbaum, Batya, and Amy Bridges. 1979. "The other side of the paycheck: monopoly capital and the structure of consumption." Pp. 190-205 in Zillah R. Eisenstein (ed.), *Capitalist Patriarchy and the Case for Socialist Feminism.* New York: Monthly Review.

Wilkening, Eugene A. 1958. "Joint decision-making in farm families as a function of status and role." *American Sociological Review* 23 (April): 187-192.

_____. 1981. "Farm families and family farming." Pp. 27-37 in R.T. Coward and W.M. Smith (eds.), *The Family in Rural Society.* Boulder, CO: Westview Press.

_____. and Lakshmi Bhradwaj. 1967. "Dimensions of aspiration, work roles and decision making of farm husbands and wives in Wisconsin." *Journal of Marriage and the Family* 29 (November): 703-711.

_____. and Lakshmi Bhradwaj. 1968. "Aspirations and task involvement as related to decision making among farm husbands and wives." *Rural Sociology* 33 (March): 30-45.

_____. and Sylvia Guerrero. 1969. "Consensus in aspirations for farm improvement and adoption of farm practices." *Rural Sociology* 34 (June): 182-196.

Wilkening, E.A. and Nancy Ahrens. 1979. "Involvement of wives in farm tasks as related to characteristics of the farm, the family and work off the farm." Paper presented at the Rural Sociological Society meetings, Burlington, VT.

Wright, Erik Olin. 1979. *Class Structure and Income Determination.* New York: Academic Press.

Part 4

The Appropriate Technology Movement

Chapter 11

Soft Tech/Hard Tech, Hi Tech/Lo Tech: A Social Movement Analysis of Appropriate Technology

Denton E. Morrison

The last decade or so has witnessed a growing general awareness of the role of technologies in our lives. Part of this stems from increasing realization that technology impacts are not all of a positive piece. There has been a lively debate over the positive and the negative character of impacts and the processes

An expanded version of this chapter appears in *Sociological Inquiry* 53, 2-3(Spring-Summer 1983). Also presented at the celebration of the decennial of the Institute for Environmental Studies, University of Illinois, Urbana-Champaign, October 1, 1982. Portions of this paper were originally in "Energy, Appropriate Technology, and International Interdependence," presented at the meeting of the Society for The Study of Social Problems, San Francisco, 1978. Parts have also been incorporated in Morrison (1980) and Lodwick and Morrison (1982). The recent work has been supported by the National Science Foundation (ISP–8016582), and earlier by the Woodrow Wilson International Center for Scholars. The Center for the Advanced Study of International Development (CASID), Michigan State University, provided support for thinking about the hi tech impacts on development. The author gratefully acknowledges these sources of support as well as the continuing support of the Michigan Agricultural Experiment Station. He also gratefully acknowledges helpful comments Dora Lodwick, Craig Harris, and Bernard Finifter made on an earlier version. This is Michigan Agricultural Experiment Station Journal Article No. 10552.

involved in producing them. Increasingly, social dimensions of technology itself and social impacts of technology have become central in this debate[1] and are clearly revealed in discussions of whether there are generic features of some technologies that make them more "appropriate" than others.

This paper describes in some detail the sociological nature of "appropriate technology" (AT) and provides a tentative and preliminary statement of issues that are likely to be part of an intense and extended societal debate on whether the rapidly emerging "high technologies" (Hi Techs) qualify as ATs.

Soft Tech/Hard Tech: The AT Movement

AT, whatever else it is or isn't, is a social movement, and it must first and fundamentally be understood as a movement. AT is a deliberate attempt to mobilize collective action to advocate and promote change, change that is regarded by those mobilized as both morally right and urgent. AT advocates hold that much, perhaps even most, current technology creates (a) inequitable social impacts, (b) impacts on the natural environment that irreversibly damage it and which lower its capacity to sustain life, and (c) impacts that in other ways decrease the quality of life. Such impacts are termed "hard" by AT advocates (i.e., harsh, difficult, unmanageable) and the technology that is claimed to create such impacts "inappropriate" or "hard technology" (HT). In turn, advocates of this viewpoint regard as imperative changing to an "appropriate" technology that will create impacts that are socially equitable, environmentally benign, and enhance the quality of life—hence the notion of "soft" (i.e., pleasant, manageable) impacts and the use of "soft technology" (ST) as a synonym for AT.

A central feature of all social movements is a challenge ideology; it contradicts predominant, established ideas.[2] An ideology is a system of beliefs and assumptions about values that gives a rationale for concrete policies, actions, and decisions; it is a guide to what is good and what is bad for society. A "challenge" ideology has two parts: (a) There is a critique of existing society and its established ideology in moral terms; i.e., people are unjustly deprived by society as it now exists. (b) There is a proposal for the changes that will rectify this; i.e., a vision of the good society and how to get there is offered.

The essence of the soft technology characterization and critique of the predominant hard technology system of production is as follows: Means of production that are capital intensive, complex, large-scale, centralized, resource-intensive and resource-exogenous have undesirable impacts. Such systems displace people, especially those in the underclass, from jobs, alienate the employed from their work, and alienate the unemployed from society. They create overabundance for a few, deprive many from their basic needs, and make the bulk of people dependent on the decisions and actions of an elite minority. They create social units that are vulnerable to external events. Such systems are destructive of the natural environment, and are thus ultimately destructive of the affluence they seek to create. Ordinary people do not participate meaningfully in such systems, but rather are mainly controlled by them.

On the other hand, productive systems that involve little capital, are small in scale, decentralized, resource-conserving, and resource-indigenous are appropriate because they have desirable impacts. They create meaningful work for all, supply the basic needs of all, promote self-sufficiency at all levels of social organization, and create an ecologically sustainable, higher quality of life. Ordinary people participate fully in such productive systems.

Terms of reference for these characterizations are international in the relations of the developed and developing countries and intranational in the relations of the overclass and the underclass.

These notions are described in greater and more comparative detail in Table 11.1, which also gives some flavor of the dialogue and debate by including the reactions of advocates of the hard technology system to the critique and proposal of soft technology advocates.

The arguments summarized in Table 11.1 make it clear that AT involves an image of society as substantially technologically determined. But the image of the process of change involved is not a simple, linear one. There is full realization, at least among the more sophisticated proponents of AT, that items of individual technology (e.g., axes and bulldozers) are important, but that they operate in a complex system of social, economic, and political institutions. This implies, of course, that both the means of production and the relations of production must be changed to achieve the soft technology society. Nevertheless, there is an implicit notion that since individual technological items are often the most immediately changeable components of the sociotechnical system, they are the logical and, indeed, the only feasible place to start.

Characteristics and Impacts of AT

The major concepts presented in Table 11.1 and the implicit theory of the change process envisioned by AT advocates are further abstracted and presented in Figure 11.1. The basic claim of AT advocates is, as indicated above, that technology with certain appropriate *characteristics* will have appropriate equity, environmental/resource, and quality of life *impacts*. The following points supplement Figure 11.1 by conveying some of the conceptual and processual fluidity of the AT notion and by highlighting other important features of the AT notion that are illustrated.

TABLE 11.1
Soft Technology (ST) and Hard Technology (HT)

The Soft Technology Characterization and Critique of Hard Technology	The Soft Technology Proposal	The Hard Technology Rationale for Itself and Its Reaction to the Soft Technology Proposal
1. Capital intensity of HT makes it unaffordable except by an economic elite.	1. Low cost, low capital, affordable technology.	1. Modern HT is expensive but also efficient and productive. ST is primitive, second-rate, inefficient, and unproductive.
2. HT involves highly concentrated capital organized in economic monocultures controlled by and benefiting mainly an elite.	2. Distributed capital, small entrepreneurship, diversity.	2. Concentrated capital and associated HT scale are economic imperatives of productive industry. ST brings cottage industry, craft economy, and associated low productivity, low incomes.
3. Large-scale economic organization of HT brings human and environmental costs that completely offset productivity benefits.	3. Technology that is small-scale, controllable, repairable, durable, reliable, safe, simple but sophisticated; small is beautiful.	3. Efficiency characteristics and productivity benefits of economies of large scale are well demonstrated; benefits clearly outweigh costs. ST is romantic, risky; small is not bountiful.

4. HT systems involve centralized, authoritarian decision making in which ordinary people have no role, no control over their fate.

5. Centralized innovation by experts creates an elite; "trickle down" of benefits tends not to reach the poor while culturally disruptive costs do.

6. Automated, large-scale HT displaces people, especially underdogs, from jobs and is exploitative and de-humanizing for the employed.

7. Specialization around HT separates, isolates, and alienates. The resulting "interdependence" is in fact hierarchical and adversarial dependence.

4. Decentralized, democratized decision making; production under the control of technology users and local communities.

5. Local innovation, development and "bottom out" diffusion of technologies related to basic needs guarantees compatibility with local cultures. No royalties, patents or import/export curbs on technological soft or hardware.

6. Employment-intensive technology, meaningful jobs.

7. De-mystification of HT by reducing scale, complexity, and segmenting of work roles. Low and fluid division of labor; low stratification in work organization. Nonadversarial relations of production. Freedom, openness and self-fulfillment in work.

4. Rational, scientific economic planning and management requires expert, centralized decision making and control. ST is economic chaos.

5. Rationally programmed R&D by experts, backed by solid basic science, has demonstrated productive merits. ST is haphazard, "tinker-tech." Development inevitably involves change toward the image of modern, developed societies.

6. HT is labor-saving. ST is backward, back-breaking.

7. Specialization in resource inputs to production, including the division of labor, is the foundation of efficiency and productivity. Workers organize to improve their situation. Conflicts are minimized through institutionalized labor-management relations. Nonspecialization around ST guarantees inefficiency and low productivity.

TABLE 11.1
Soft Technology (ST) and Hard Technology (HT)
(continued)

The Soft Technology Characterization and Critique of Hard Technology	The Soft Technology Proposal	The Hard Technology Rationale for Itself and Its Reaction to the Soft Technology Proposal
8. Work in HT society is for greeds, not needs, i.e. consumer "demands" artificially created to sustain "growth" in an economy involving complex, fragile interdependencies.	8. Emphasis on local production for sustained provision of local basic needs. Emphasis on quality of life, not economic growth.	8. Only through demands for goods created in affluent, developed sectors will jobs be created and basic needs satisfied in the developing, poorer sectors. Economic growth benefits all as the whole ladder rises. ST involves leveling at an unacceptably low level.
9. HT creates simultaneous consumer affluence for a few and high unemployment.	9. Full employment, moderate levels of living, minimal income difference.	9. Affluence is intrinsically good, and moreover provides the surplus to allow even the unemployed to have basic needs provided. Full employment at a subsistence level with ST has no intrinsic merit.

10. HT wastes, depletes, and denigrates natural resources and creates dangerous environmental pollution. HT creates competition for increasingly scarce resources and the associated means for destruction via weaponry.	10. Sparing use of indigenous natural resources, especially nonrenewable ones, reduces the potential conflicts over scarcities. Ecologically sound development that preserves environmental quality.	10. Rational management, development and utilization of natural resources, including a continuing search for substitutes for scarce resources, brings sustained productivity. The "limited good" mentality of ST means economic underdevelopment and stagnation. Only *more* HT can preserve environmental quality, or security.
11. HT moves society toward large, highly centralized, urbanized communities that are crowded, interdependent, vulnerable; the countryside is depopulated.	11. Decentralized, small, self-sufficient, relatively autonomous communities where production and consumption are integrated around basic need provision through ST at the local level, thus reducing vulnerability to outside actors and increasing social resiliency.	11. Concentrated populations have well known advantages in increasing the levels of economic, social, and cultural amenities. The ST notion of community self-sufficiency is romantic; rural communities are inevitably backward, provincial, isolated and poor.

Characteristics Interrelationships (Figure 11.1, "X")

The characteristics of technology of concern in AT can be classified, as in Figure 11.1, into "basic," "environmental/resource," and "processes" categories, a codification that is somewhat arbitrary. These categories and subcategories are intimately interrelated in the minds of AT advocates. For instance, hydroelectric facilities that are large-scale are a central source of electricity, involve complex technology, and are nonparticipatory in their management and operation for the users of the electricity. A small-scale hydroelectric facility is less centralized in the sense of serving a smaller geographic and population scope, less complex, and more amenable to local ownership as well as construction, management, operation, and repair by local talent. Also, the basic characteristics of concern in AT often merge hardware and organizational dimensions of technology. For instance, the construction and operation of a major hydroelectric facility would involve specialized equipment *and* a specialized work force for installation, operation, management and maintenance, with impacts on that work force as well as on the larger society from which that work force is selectively recruited, trained, and rewarded.

Characteristics and Impacts Interrelationships (Figure 11.1, "Y")

Some of the soft technology "characteristics" are themselves desired "impacts" and all the "characteristics" are claimed to be the causes of other desired "impacts." For instance, participation in technological decisions is itself an intrinsically desired impact in the soft technology notion, but participation is also a part of the process of insuring that technologies produce employment and address basic human needs. The quality of life is viewed as inseparably related to these features of the productive system.

For analytic purposes a distinction can be made between the "characteristics" and "impacts" of technology, but these features blend almost indistinguishably in AT advocacy. Additionally, AT advocates do not necessarily see a one-way causal path between the characteristics and impacts of technology. Although the characteristics of technology are viewed as the main independent variables, AT places strong emphasis on the necessity of value change and the causal role of value change, e.g., it is thought that an increased concern for equity is crucial *and* will result in smaller scale technologies. Fully as much AT advocacy argues for increased priority for equity and the other values implied in the impact sub-categories as argues for particular claims about the technological changes needed to move toward such values. The arguments for values tend to be indistinguishable from the arguments for the claims. Similarly, debates between soft and hard technology advocates often tend to involve unrecognized, implicit, and probably irresolvable differences in value preferences. The value preferences become tangled in disputed—even though seemingly empirically

FIGURE 11.1

Major Appropriate Technology Characteristics, Impacts, and Interrelationships

Characteristics	Impacts

Basic
- Small scale
- Decentralized
- Simple
- Nonspecialized
- Light capital
- Labor intensive

Equity
- Low Cost
- Employment enhancing
- Basic human need provision
- Participatory
- Self-reliance enhancing
- Egalitarian

Environmental/Resource
- Nearby
- Renewable
- Sparing use, recycling

Environmental/Resource
- Ecosystem sustaining
- Environmentally benign

$$X \qquad Y \qquad Z$$

Processes
- Participation in:
  - Knowledge
  - Innovation
  - Construction
  - Operation
  - Repair
  - Management
  - Decisions
  - Control
  - Ownership
- Incremental
- Locally engendered/oriented

Quality of Life
- Nonalienating
- Humane
- Controllable, comprehensible
- Culturally compatible

resolvable — claims about their relationship to basic features of technology, the resources involved, and the process of change.

Impacts Interrelationships (Figure 11.1, "Z")

Important interconnections among the impact categories and sub-categories become apparent when concrete technologies are considered. For instance, wood burning for cooking and heating in open fires brings deforestation and soil erosion. These impacts, in turn, are, over time, related to deprivation from basic human needs. It is claimed that simple mud stoves (e.g., Lorena stoves) made locally will burn wood more efficiently and thus have positive environmental impacts at the same time that they improve equity for women and improve the quality of life for women and children by reducing time spent in wood gathering. Alternatives such as kerosene stoves are hard technologies because they are not within the means of most and also decrease local self-sufficiency.

The example of the Lorena stove makes clear that certain environmental/ resource and quality of life impacts are neither irrelevant to nor easily separated from equity impacts. The example is instructive, particularly since it shows these interrelationships in a developing country technology. But it should not be allowed to obscure an important distinction about the impact categories; namely, that the three broad classes of impacts differentiate the main concerns of the developed and developing country tributaries that merge to form the international AT movement. There is no basis for exclusive foci, as the example shows, but equity impacts are the predominant concern in the developing countries, while environment/resources and quality of life impacts are the central concerns in the part of the movement that advocates AT in the developed countries. Many social movement organizations in the developed countries have "outposts" in the developing countries which feature equity as their main concern. All the groups that have the advocacy of AT in and for the developed countries as their focus acknowledge the legitimacy of the special emphasis on equity impacts for AT in developing countries and most of these groups give high priority, although not generally the highest priority, to equity impacts in their own context.

The Hi Techs

Interest in AT grew steadily in academic, public interest, and policy circles after the mid-1970s. This interest was spurred by an increasingly coalesced, viable, and independent AT movement, the concerns the movement articulated, and the way these concerns were amplified by allies in related movements and elsewhere. But the AT notion never achieved the currency of high technology, another characterization of technological style that emerged in roughly the same timeframe. "Hi Tech" rapidly became part of media jargon and, in turn, a part of popular language after its birth in the marriage of advanced science and entre-preneurship. Hi Tech has also found a prominent place in the current vocabulary of public policy.

The Hi Tech notion tends to be popularly associated with sectors of the economy characterized by their type and rate of innovation, especially aero-space, computers, and communications. But the Hi Techs are not by any means restricted to these sectors. The term refers more basically to one or more of the following types of innovations:

- *Electronic technologies* for the creation, storage, retrieval, transmission, manipulation, and control of information, including information for the automated control of mechanical systems.
- *Biological and biochemical technologies* related to medicine, industrial processing and material fabrication, agriculture, food processing, and energy conversion.
- *Materials technologies* to achieve resource input efficiencies and/or to achieve the strength, weight, durability and size characteristics needed for traditional (e.g., goods, transport, shelter, clothing) and nontraditional (e.g., space, tissue, deep seabed, extreme cold and heat) applications.

The essential feature of Hi Tech development is advanced scientific knowl-edge, both basic and applied. A good bit of Hi Tech involves instrumentation for basic and applied research and the development of technologies originally used in research for broader commercial and consumer applications. The driv-ing social force behind Hi Tech is economic: efficiency, productivity, and profita-bility in the creation and marketing of innovative, competitive goods, services, and processes.

The Hi Techs modify the organization of traditional industry, but often without altering basic patterns, e.g., robots in auto assembly, automated type-setting, new herbicides for new plant varieties in agriculture. But the growing magnitude of Hi Tech and the special character of its knowledge requirements increasingly creates a new productive sector with an associated organizational infrastructure. Within the context of strong proprietary concern for the technol-ogies, this infrastructure is designed to attract personnel and to accelerate and otherwise facilitate interaction of the needed basic and applied science inputs, as well as to transmit the new knowledge and skills associated with making, selling, and using the Hi Techs, e.g., "parks" for electronic and biotech research, devel-opment, production, and marketing around universities. Krimsky (chapter 3 in this volume) provides further details on the extent of these developments.

The Hi Techs and Soft Impacts

The current interest and activity in Hi Tech is intense; it includes a strong sense of urgency. Although certain moral overtones may be involved in the advocacy of Hi Tech, it is mostly the urgency of returning a sagging, inflationary economy to a path of growth and general, high affluence, combined with the conviction of some that increased emphasis on Hi Tech is necessary to accomplish this. Clearly, Hi Tech emerged within and is firmly embedded as a part of the economic growth paradigms, both capitalist and socialist, but especially the capitalist.

This does not mean that the Hi Techs are completely incompatible with the AT notion. Neither does it mean that the actual and potential sources of support for AT will be uniformly negative in their reactions to all aspects of Hi Tech.

The following analysis examines Hi Tech mainly in terms of the three major types of impacts of concern in AT (equity, environmental/resources, quality of life) and introduces, as needed, some of the other major notions associated with AT (see Figure 11.1). The discussion is intended to be preliminary, speculative, and heuristic. It is aimed at generating thought and debate that will help specify some of the major issues for further, closer theoretical and empirical examination. Examining Hi Tech from an AT perspective reveals many uncertainties about Hi Tech *and* about AT impacts. But AT provides a reasonably coherent, explicit conceptual framework for examining the Hi Techs, a framework that focuses on social dimensions and social impacts. The following analysis is, however, only a small and specific slice of what has emerged and will continue as a broad and pervasive concern (Toffler, 1980; Nora and Minc, 1981).

Equity Impacts

An important source of the interest in the productivity and efficiency characteristics of Hi Tech is the desire to reduce the cost of productive processes, including, importantly, labor costs. Since increased employment opportunities and the associated stress on productive processes that are labor intensive are central features of the AT concern for equity, there is a fundamental sense in which the Hi Techs have hard impacts. The situation is, however, more complex than this; much more complex than can be indicated in this brief sketch.

Ironically, the press for equity by various movements of the underclass (e.g., organized labor) has been one of the factors in driving up the cost of labor inputs. This press has taken place in parallel with a powerful environmental movement that has independently had the effect of increasing the cost of other (natural resource related) inputs. Further, the environmental reforms have had a generally constraining influence on economic growth – the traditional way of addressing equity claims in the capitalist paradigm.

The concrete, contractual expression of equity concerns in organized labor has often meant that efficiency concerns pushed owners to introduce labor-saving technology, thereby reducing the number of workers. But the remaining workers often have received improved wages, benefits, and job security. Thus, the equity issue has become somewhat blurred for the underclass. Also, it is often claimed by Hi Tech advocates that the jobs eliminated are the most undesirable ones from the standpoint of tedium or occupational health, e.g., replacement of routine assembly work by robots, or automated handling of toxic materials. Further, according to Hi Tech advocates, even if the Hi Techs reduce jobs in particular (usually traditional) industries, the net, long-term impact of Hi Tech will be to increase employment opportunities by creating new products, new services, and new demands. The broadest version of this claim is that the Hi Techs will stimulate economic activity generally, contribute to economic growth, and thereby have salutary equity impacts (i.e., in the capitalist notion of equity).

Environmental/Resource Impacts

Advocates claim that the efficiency characteristics of Hi Tech extend to the quantity and types of resources used in productive processes as well as to uses that are more environmentally benign. And there is, in fact, little doubt that the OPEC action and the continuing, substantial concern with environmental impacts put a considerable focus on energy efficiency. It has become standard by now to interpret energy conservation as being best achieved by increased energy efficiency — getting more output from the same energy input — rather than a reduction in the functions energy performs or in the amenities energy provides. This has been merged, in theory and practice, with the search for energy alternatives, including the soft alternatives. The Hi Techs have been an important part of this, e.g., "smart" thermostats and other energy control devices, new materials for insulation and solar heat collection, photovoltaics, and so on. A major dilemma for those concerned with equity is that higher prices for conventional energy are regressive and, similarly, the first-costs associated with many Hi Tech energy conservation and alternative energy sources have regressive impacts (Morrison, 1978). These costs make such Hi Techs of questionable appropriateness, and some of the Hi Techs associated with certain energy alternatives are viewed by AT advocates as undesirable on environmental/resource *and* equity grounds, e.g., nuclear energy and solar satellites.

There is a strong advocacy within the environmental movement and elsewhere for energy technologies that are simple, use local skills and local, renewable resources. This strategy thus addresses environmental/resource problems while simultaneously contributing to equity by their labor intensivity, low cost, and by focusing on basic human needs (Morrison, 1980). The favor given to such technologies by environmentalists is, in part, a reaction to equity dilemmas posed by some of the Hi Techs for energy, but the latter are still generally seen by environmentalists as appropriate for applications where energy efficiency is the central concern. For instance, Lovins clearly favors Hi Techs for industrial applications in developed contexts (1976, 1977, 1978).

Some of the Hi Techs, moreover, promise soft environmental *and* equity impacts simultaneously. For instance, some of the biotechs promise the following:

rapid reforestation; efficient biogas production from crop residues, animal, and human wastes; microbiological processes for production of substitutes for petrochemical feedstocks from wastes or from other abundant, renewable materials; sludge digestion to produce fertilizers; drought and pest resistant plants for agricultural use; nutritional and medical improvements for humans and farm animals; and so on.

Biotechs of these sorts and others have important potential applications in developed *and* developing countries. Those that promise to produce energy for local community and household applications or that promise improved agricultural productivity have a special salience for the critical problems of the poor in developing countries. To date, however, Hi Tech activities, including biotechnology research and development, have not, in the main, focused on applications for the developing countries, nor on applications to basic human needs in developing or developed contexts.

Quality of Life Impacts

Two more or less diametrically opposed views on the quality of life impacts of the Hi Techs are commonly projected, especially for the electronic technologies (Coates, 1982; Renfro, 1982).

The positive view asserts that the Hi Techs will humanize work by decentralizing it, e.g., less separation of home and work, less need for crowded urban agglomerations, less time and energy spent in commuting, shopping, etc. —the "electronic cottage" notion (Deken, 1982). Individuals will be better able to control their relationship to work in terms of timing and pacing and also to better control their relationships to machines, peers, and supervisors. The organization of work will be less hierarchical, more participative, and less controllable by managers since work will be less dependent on locational characteristics, e.g., automated factory processes could be controlled from terminals in homes.

There will be more freedom, autonomy, and opportunity for the creative integration of work, play, education, and family activity. Sophisticated communication devices will promote democratic community participation by allowing instant, effortless registry of opinions on alternative proposals. Enormously greater access to information will increase the opportunity for self-fulfillment through learning, as well as facilitate decision-making and everyday problem-solving.

The negative view asserts that automated workplaces will not only reduce employment opportunities, robots will continue the trend toward total elimination of skilled craftswork and the rich human relationships and material mastery involved in the process by which crafts and skills are transmitted and executed. Work may be more decentralized in a spatial sense, but the content of work and the process of management and monitoring will require a high degree of coordination by a centralized management. It will involve an increased dependence on specialized experts for the manufacture, installation, and servicing of devices for productive activities as well as for consumption. The level of interdependence involved will not promote autonomy and self-reliance for individuals or for communities. Both products and processes will be increasingly standardized, even if the timing and setting of work introduces flexibility. Computer augmented rational-choice using strictly economic criteria will govern all entrepreneurial decisions, and these decisions, centrally made, will increasingly be paramount in determining citizen welfare.

The electronic aspects of marketing will insure that it is mass marketing and will involve only the fine-tuning of long-term trends in the artificial, large-scale creation of needs for goods and services that are, in the first instance, profitable. Freedom and diversity of choice by electronic processes will be largely illusions created by and maintained by electronically enhanced media hype to manipulate consumers. Electronic participation in decision processes will be manipulated by central sources in the same way.

Contrasting positive and negative views one notes that perhaps even more important than the specific quality of life impacts projected from Hi Tech in these two scenarios are the equity overtones and implications. Expanding somewhat on the negative scenario, the Hi Tech critics fear that an elite layer of experts will have more or less exclusive access to the new specialized knowledge involved in the invention, construction, servicing, and use of the Hi Techs and, consequently, this layer will capture most of the rewards. Quality of life will increase for this strata of society but only if it is defined in terms that emphasize a lifestyle of electronically enhanced consumer affluence, with a particular emphasis on information access, rather than one with concern for widespread, increased self-reliance in the provision of basic human needs. It is thus strongly implied that the intra- and international layers of the stratification system will be further spread and solidified. Domestically, the special role of technical subcultures, networks,and associated language and conceptual features will tend to give a special importance to technical education. Intergenerational aspects of stratification are likely to be exacerbated. In short, stratification will increasingly be based on technical knowledge and information and on access to the associated means of production as a basis for material reward, life chances, prestige, and political power. There is, in fact, some fear that the electronic technologies will dichotomize society into "information rich" and the "information poor."

Admittedly there are some features of the positive scenario that are consonant with the AT view (Burns, 1981). But the negative perspective is here posited to reflect the predominant reaction of AT advocates to the quality of life impact potential of the Hi Techs. Indeed, this view reflects much of what AT advocates will find objectionable about the prospects of a Hi Tech future in general. This is because equity impacts are central in AT; negative equity and negative quality of life impacts virtually become one and the same in consideration of the Hi Techs.[3]

Summary and Conclusions

The terms "appropriate" and "technology" have meanings in the ordinary language; it may seem reasonable that the meaning of their conjunction in "appropriate technology" can be easily derived by logical/common sense considerations. But this is not the case. The meaning of AT is given — albeit loosely — by the AT movement.

AT is a set of concerns about impacts and an associated set of notions about sociotechnical features that are claimed to produce these impacts. The impacts can be broadly categorized as (a) equity, (b) quality of life, and (c) environmental/resource. The equity impacts are the most fundamental. The concern for equity pervades and shapes the nature of the concerns for the other impacts. It is for this reason that the AT notion has been attractive to environmentalists in addressing the common accusation that they are elitist.

AT can also be characterized as an alternative paradigm of development. The AT paradigm envisions, in contrast with traditional socialist and capitalist paradigms, a de-escalation of growth and general affluence in developed sectors and a stable merging of the poor and the rich at a level somewhat above basic human need provision.

The Hi Techs do not fit this alternative paradigm nearly as well as they fit the more traditional growth paradigms, especially the capitalist. The Hi Techs have their roots in advanced science and are addressed mainly to the growth needs of the developed countries.

There may, however, be some basis for uncertainty about a general assertion that the Hi Techs do not basically fit the AT paradigm. Certain of the Hi Techs could, if focused on developing countries, address their equity and environmental/resource problems. This is particularly a potential of biotechnology. Also, since one of the signal features of Hi Tech is facilitation of rapid, accurate information flow without special geographic limitation, it is possible to envision a scenario involving extension of the decentralization impacts of Hi Tech well beyond the boundaries of the developed countries. There is, for example, no theoretical reason that auto assembly robots located in Detroit could not be programmed or their output, properly measured by Hi Tech devices, monitored and controlled from computer terminals in Lagos via satellite connections.

Obviously such a notion assumes a labor force in Lagos that is more cheaply trained, outfitted, managed and paid than the one available in Detroit, with all that this implies for domestic equity problems. It also implies that the relevant training can be transferred to a labor force that is not as "modern" as the one available in Detroit. Perhaps, however, the Hi Tech potential for developing countries is the most optimistic in this regard, for the knowledge and training associated with Hi Techs, given the sophisticated communications that are part of Hi Tech, may be more easily transferred to developing countries than some of the requisite features of Lo Tech. Within certain limits, the Hi Techs are less dependent than the Lo Techs on resources in specific locations, e.g., mineral deposits, land, water, machinery, factories, transportation infrastructure, etc. It may be that even the means for research and development are, with Hi Tech, less dependent on traditional physical and personnel infrastructures in particular, traditional geographic locations, e.g., the Hi Tech parks of the future may be as much electronic in their physical and locational dimension as in the substantive direction of their research and development efforts.

Still, it is very questionable whether the Hi Techs can be successfully transferred to contexts where a developed society infrastructure is not fully and firmly in place. For instance, the electronics technologies require a large and completely reliable supply of electricity, a full inventory of highly specialized parts plus the associated repair personnel and organizations, etc. Some of the biotechnologies for agriculture require an infrastructure of sophisticated hardware and personnel for fine-tuned monitoring of weather conditions, soil moisture content, pests, and so on.

Beyond the proper labor, material, and other support infrastructure, the transfer of Hi Tech to developing countries would probably require a considerable extension of current multi-national organizations, plus an even greater concentration of capital in these endeavors than exists currently. It can be guaranteed that these and other features of Hi Tech will be greeted with something much less than enthusiasm by AT advocates. The proponents of Hi Tech will argue that some negative impacts are a minor and, at any rate, inevitable part of what should be seen as reasonable short-term tradeoff for developing countries. The long-term payoff, it will be argued, is a technologically sophisticated labor force, eventual autonomy in Hi Tech research and development, capital accumulation, economic growth—and all it brings.

This argument will be familiar to AT advocates. They will point out that the general failure of such impacts to occur in the past, and especially the fact that this development paradigm did nothing for massive rural poverty, is what led Schumacher and his followers to the AT notion in the first place.

There is, then, on the whole, little basis for belief that AT advocates, especially those who focus on international development, will see much reason for thinking of Hi Tech as having soft impacts. On the contrary, they will see Hi Tech as the cutting edge of hard technology.

But, ready or not, like it or not, the Hi Techs are here—or at least approaching with great momentum. An obvious part of the reason the Hi Tech notion has so rapidly become an integral part of popular language while the AT notion is still sub-cultural is the fact that "Hi Tech" refers to things that are happening all around us as a part of our everyday lives. Hi tech is a clearly discernable direction of change in society.

The Hi Techs are being deployed in a context of high unemployment, high inflation and a generally faltering global economy. Domestically, these problems are severe and there is, in addition, wide perception and fear that hard-won environmental reforms, as well as programs designed to add security and dignity to the lives of the poor and the elderly, are under great threat. The stage is set for the revitalization of movements for equity, environment, and quality of life. In fact, we see the predictable actors starting to play their predictable roles.

The AT movement may not be at the visible vanguard of the intense dialog and social conflict that will ensue over these issues, but it will be an active, vital direct and indirect force. Because of the AT movement it is improbable that these issues will in the future be debated as separate and unrelated concerns for separate and unrelated movements. The influence of the AT movement also makes it improbable that the debates can avoid consideration of the larger, even global, context, despite the fact that the debates will mostly take place in particular geographical settings, often local ones. *Thinking Globally, Acting Locally* (Feather, 1980) is AT in a nutshell.

The AT movement has lengthened the menu of technology, but, more important, it has both broadened and sharpened the moral, conceptual, and geopolitical criteria that will, in the future, be important in technology choice. Domestically we now have official advocacy of the "New Federalism." It remains to be seen whether or not it is compatible with the "New Globalism" (or should it be the "New Glocalism?") of AT advocacy.

Notes

1. The scholarly concern for the technology-society relationship is, of course, a long-term one. Spiegel-Roesing and de Solla Price (1977) and Durbin (1980) document much of the intellectual history of this concern and provide extensive bibliography.
2. The basic source of the AT ideology is, of course, Schumacher (1973), but many others have amplified and extended Schumacher's arguments in important ways. McClaughry (1978) has listed the central general items and Borremans (1979) presents an expanded, annotated list which includes much that is technical. There is, additionally, an analytic literature specifically focused on AT, e.g., Jequier (1976), Dickson (1974), Robinson (1979), Norman (1978), Long and Oleson (1980), Stewart (1977), Diwan and Livingston (1979) and Rybczynski (1980). In addition, much of the vast general technology-and-society literature mentioned above is relevant to the analysis of AT.
3. The tendency for opposite projections of Hi Tech impacts also occurs where equity is considered alone. See Irma Adelman's dissenting view on "Remote Sensing and Equity" in the generally positive report of the Committee on Remote Sensing for Development (1977). For a recent, sensitive (and in part empirically based) analysis of the equity-related domestic impacts of the electronic Hi Techs on public decision making, including a discussion of some ways such technologies might be implemented to avoid negative impacts, see Danzinger, *et al.* (1982).

References

Borremans, Valentina. 1979. *Guide to Convivial Tools.* New York: R.R. Bowker.

Burns, Alan. 1981. *The Microchip: Appropriate or Inappropriate Technology?* Chichester, England: Ellis Horwood, Limited.

Coates, Joseph F. 1982. "Telematics, demography, and the end of alienation." *National Forum* LXII (Spring):26–88.

Committee on Remote Sensing for Development. 1977. *Resource Sensing From Space: Prospects for Developing Countries.* Washington, D.C.: National Academy of Sciences.

Danzinger, James N.; William H. Dutton; Rob Kling; and Kenneth L. Kraemer. 1982. *Computers and Politics: High Technology in American Local Governments.* New York: Columbia University Press.

Deken, Joseph. 1982. *The Electronic Cottage.* New York: William Morrow.

Dickson, David. 1974. *The Politics of Alternative Technology.* New York: Universe Books.

Diwan, Romesh, and Dennis Livingston. 1979. *Alternative Development Strategies and Appropriate Technology: Science Policy for an Equitable World Order.* New York: Pergamon.

Durbin, Paul (ed.) 1980. *A Guide to the Culture of Science, Technology, and Medicine.* New York: The Free Press.

Feather, Frank. 1980. *Through the '80s: Thinking Globally, Acting Locally.* Washington, D.C.: World Future Society.

Jequier, Nicolas (ed.) 1976. *Appropriate Technology: Problems and Prospects.* Paris: Development Centre, Organization for Economic Cooperation and Development.

Lodwick, Dora G., and Denton E. Morrison. 1982. "Appropriate technology." Pp. 44–53 in Don A. Dillman and Daryl J. Hobbs (eds.), *Rural Society in the U.S.: Issues for the 1980s.* Boulder, CO: Westview Press.

Long, Franklin A. and Alexandra Oleson (eds.). 1980. *Appropriate Technology and Social Values: A Critical Approach.* Cambridge, MA: Ballinger.

Lovins, Amory B. 1976. "Energy strategy: the road not taken?" *Foreign Affairs* 55 (October):65–96.

_____. 1977. *Soft Energy Paths.* New York: Harper.

_____. 1978. "Soft energy technologies." *Annual Review of Energy* 3:477–517.

McClaughry, John. 1978. *A Decentralist Bookshelf.* Concord, VT: Institute for Liberty and Community.

Morrison, Denton E. 1978. "Equity impacts of some major energy alternatives." Pp. 164–193 in Seymour Warkov (ed.), *Energy Policy in the United States: Social and Behavioral Dimensions.* New York: Praeger.

_____. 1980. "The soft, cutting edge of environmentalism: why and how the appropriate technology notion is changing the movement." *Natural Resources Journal* 20(April):275–298.

Nora, Simon, and Alain Minc. 1981. *The Computerization of Society: A Report to the President of France.* Cambridge, MA: MIT Press.

Norman, Collin. 1978. *Soft Technologies and Hard Choices.* Washington, D.C.: Worldwatch Institute.

Renfro, William L. 1982. "Second thoughts on moving the office home." *The Futurist* 16(June):43–48.

Robinson, Austin (ed.) 1979. *Appropriate Technologies for Third World Development.* New York: St. Martins Press.

Rybczynski, Witold. 1980. *Paper Heroes: A Review of Appropriate Technology.* Garden City, New York: Anchor Books.

Schumacher, E.F. 1973. *Small Is Beautiful: Economics As If People Mattered.* New York: Perennial Library.

Spiegel-Roesing, Ina, and Derek de Solla Price (eds.) 1977. *Science, Technology and Society: A Cross Disciplinary Perspective.* Beverly Hills, CA: Sage.

Stewart, Frances. 1977. *Technology and Underdevelopment.* London: Macmillan.

Toffler, Alvin. 1980. *The Third Wave.* New York: William Morrow.

Chapter 12

Soft Energy and Hard Labor? Structural Restraints on the Transition to Appropriate Technology

Allan Schnaiberg

Introduction

In the winter of 1973, a dramatic new book by the distinguished European econo-mist, E.F. Schumacher, appeared on the American political scene: *Small Is Beautiful: Economics As If People Mattered*. If one were to arrange a social marketing approach for this work, it could certainly not have been more success-ful than the actual context Schumacher's work emerged in. As a humanistically-oriented work, it found much favor with the critics of neo-classical economics, whether from the liberal critics who decried the absence of an understanding of economic institutions, or from the neo-Marxists and related political economists who decried the separation of technical economic decision-making from the poli-tics that contaminated it in the real world (e.g., Dickson, 1974: ch. VI). More-over, it took up a humanistic concern with the problems posed by the environ-mental movement of the late 1960's and especially with the difficulties of legislat-ing and enforcing environmental protection in economies directed by profit motivations rather than the public good (Morrison, 1980; *cf.* Mitchell, 1980).

But perhaps best of all, the book emerged in the winter of the first "energy crisis," provoked by the Organization of Petroleum Exporting Countries (OPEC) boycott of western markets, and the book was marketed throughout the

The author wishes to thank Denton Morrison for his extensive comments on a previous draft of this chapter.

succession of later price increases stemming from the initial success of this unprecedented boycott by this oligopsonistic group of oil producers. This challenge to availability and cost of one of the key factors of western (and eastern) industrial production caught American political and economic elites largely unprepared, with jingoistic rather than strategic political and economic changes as their initial responses. Schumacher's work found favor with a disparate group of social movement advocates and political figures alike, because it offered one cogent alternative—social, political, and economic—to this challenge by "foreign powers" (Schnaiberg, 1983a). His solution was elegantly simple: we could opt out of the competition for huge reserves of foreign oil by reconstructing our economy around "appropriate technology," which delivered social utilities at a far lower energy, material, and capital cost. Rather than our perpetual search to solve our social needs by "inappropriate technology," we would find a permanent and peaceful solution in a socially appropriate technology. Our salvation lay in conceiving of and implementing a whole series of changes in the physical technology to be incorporated into societal production, involving human access to and involvement in direct production, and concomitant reductions in scale and unnecessary complexity of production and distribution (Morrison, 1978, ch. 11 in this volume). The ultimate mix of technology would be "intermediate" between the primitive technology of pre-industrialism, and the advanced technology of modern industrialism.

Some three years later, a second and briefer work appeared, by the physicist Amory Lovins: "Energy Strategy: The Road Not Taken?", in the prestigious *Foreign Affairs* journal. Although not directly linked with Schumacher's work, it strongly reinforced Schumacher's political-economic premises with an energy calculus and some basic economic data arguing for a decentralized energy system in the U.S., with energy sources more closely tied to the social end uses of the energy generated. In the following year (1977), Lovins elaborated his argument in the influential work, *Soft Energy Paths: Towards a Durable Peace*, claiming that his form of "appropriate technology" (AT)—the "soft energy technology"—would, in the context of changed social values oriented toward peace and permanence, also help redirect social production and organization toward a "soft energy path" (SEP) that would reduce social production and maintain the "durable peace" proclaimed in the title (Morrison and Lodwick, 1981; Morrison, ch. 11 in this volume.)

As with Schumacher, the timing of Lovins' works could not have been more propitious. In the 1976–1977 period, legislative and administrative efforts to enhance energy production were frequently defeated, and the beginnings of a coalition to support energy conservation through direct incentives and disincentives was established, with some initial support of President Carter. In the midst of these legislative dead-ends, public outcries for a coherent energy policy rose, as impacts of increased prices of energy slowly grew, and as perceived vulnerability of American oil supply also grew. Demand for policy-making, concurrent with temporary and partial repudiation of expanded "hard" energy technology —breeder reactors, synthetic fuels, and unregulated coal production—generated greater receptivity to "new" approaches such as Lovins'[1]. Both Schumacher's and Lovins' works offered a *positive* option to mere critiques of past energy and social production (Morrison, 1978, 1980), then, and an option to a simple management of scarcity (Schnaiberg, 1980:425).

Some Theoretical Problematics of Appropriate Technology Programs

As attractive as these proposals of Schumacher and Lovins have been to the environmental movement (Morrison, 1980)—and some related social equity movements, such as the neighborhood and poverty movements (e.g., Primack, 1980; Pitts, 1981)—there are some major pitfalls that have not been recognized by most of the social science community. These will be treated at the theoretical level in this section, and illustrated with the trajectory of American energy conservation policy and practice, in the following ones. While the focus in this chapter is on energy issues and urban-industrial production, the analysis has applicability in both rural and urban settings (Hackett and Schwartz, 1980), and for agricultural and industrial production (Morrison, ch. 11 in this volume). Likewise, the macrostructural issues have counterparts at microstructural levels, ranging from small groups to communities in both rural and urban settings (e.g., Hackett and Schwartz, 1980; Pitts, 1981; Primack, 1980). The focus on energy also relates to food and other agricultural production, since agriculture both generates and consumes much of society's energy, including energy directly consumed by the population (Pimental, *et al.*, 1973; Commoner, 1977).

The Schumacher and Lovins models of social change through technological innovation can be charted, as in Figure 12.1.

FIGURE 12.1

Simplified Translations of Schumacher and Lovins Models

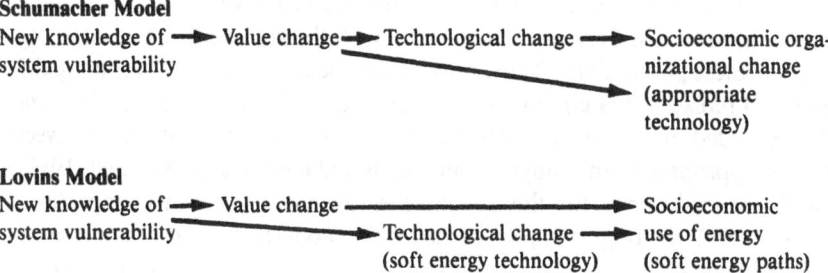

Schumacher Model
New knowledge of → Value change → Technological change → Socioeconomic organizational change (appropriate technology)
system vulnerability

Lovins Model
New knowledge of → Value change → Socioeconomic use of energy (soft energy paths)
system vulnerability → Technological change (soft energy technology)

From Figure 12.1 the major differentiation of Lovins and Schumacher is in the rather more direct role of *knowledge* influencing technological innovation or transfer in Lovins' work, and the rather more central role of *value* changes in Schumacher's. This is consistent with the popular view of Lovins as more technically-oriented and Schumacher as more philosophically-oriented.

Schumacher believes that the "problem of production" has not been "solved," since there continue to be excessive social and resource costs in modern industrialization. Lovins recognizes that "hard energy paths" are wasteful of both energy and human potential. Both thus seek to reshape production. They and the movements they have sponsored have an extensive critique of the modern "treadmill of production" (Schnaiberg, 1980: ch. V). Morrison (1978, ch. 11 in this volume: Table 11.1) provides the most succinct summary of this critique. Neither author, however, deals systematically with the institutional questions that these challenges raise: e.g., (1) *who* recognizes these limits? (2) what *power* do these individuals and groups have to disseminate these ideas? (3) conversely, what groups *resist* either this recognition and/or this dissemination of knowledge? and (4) how is this struggle between *contending* consciousness factions to be resolved? (Frahm and Buttel, 1980; Dickson, 1974; Cooper, 1973).

In contrast to this conflict perspective, Schumacher and Lovins and their followers frequently seem to implicitly follow the predictions of diffusion-of-innovation models of technological change (e.g., Leonard-Barton, 1980; Leonard-Barton and Rogers, 1981; Fliegel and van Es, ch. 1 in this volume). Social transformation is linear and complete, since new ideas replace old. Once initial "error" is recognized in the social production system it will be succeeded by "correctness" in ideas, values, and behaviors. This is, indeed, as Schumpeter (1947: 200–205) once so elegantly put it, theory about "archangels" succeeding their predecessors who had lesser insights: newcomers now have perfect insight and there is perfect succession of this insight. This is totally unrealistic even for socialist revolutionary change (e.g., Goldman, 1972, Enloe, 1975).

Several things negate this implicit diffusionist model, however, particularly in advanced industrial countries with high levels of material welfare (Morrison, 1978). First, many of the technics favored by appropriate technologists like Schumacher are variants of older technics that have been rejected; thus we should expect *resistance* to diffusion arising from a variety of sources. Second, the goals of voluntary simplicity advocated by many appropriate technologists (e.g., Hackett and Schwartz, 1980) are in direct conflict with both the goals of modern industrial society populations, and with the achievements of large segments of these populations. Thus, rather than "postmaterialism" becoming diffused and replacing materialism (e.g., Inglehart, 1977; Marsh, 1975; Watts and Wandesforde-Smith, 1980), we should anticipate an ongoing conflict between these "inappropriate" and "appropriate" goals and means (e.g., Mitchell, 1980). This is especially true for the Schumacher agenda for social change (e.g., Cooper, 1973), although it applies to the Lovins model as well, insofar as the latter is an agenda for soft energy *paths* and not merely soft energy *technologies*.

Once we place AT or SEP in a conflict perspective, then the central sociological questions about these models become: (1) what are the comparative
strengths of the contending forces? (2) what changes in these relative strengths
are likely to occur? and (3) what will be the future outcome of these changes?
(e.g., Buttel and Larson, 1980). The guiding assumption of the AT–SEP theory
is the initial strength of the contending forces is irrelevant, since the "facts" will
soon persuade enough (or all) of the unnamed opponents to accept the necessity
for change of values and behavior. It is thus a theory of functionalist form, in
which the need for personal and social survival under limited resources will
dictate one true path – AT and/or SEP – at some unspecified rate of change.
Moreover, changes in the forces of production *will* also happen and *will* produce
progressive egalitarian changes in the relations of production; this theory is thus
another form of technological determinism (Etzkowitz, 1981; Morrison and
Lodwick, 1981; Morrison, 1980, ch. 11 in this volume). Stretton (1976), among
others, instead indicates a *range* of sociotechnical changes which could occur
because of resource limits – many of which are socially regressive. Nothing in
the Schumacher-Lovins theory speaks against the likelihood of these changes
occurring, except that social *values* will change sufficiently to preclude this.

But social history is replete with *aggregate* social risks – resource limits
among them – being highly *unequally* shared. Indeed, without this history sociologists would have had little enduring interest in social stratification and social
inequality. Moreover, this history itself helps us understand not only the likelihood of *alternative goals* (values) arising in our confrontation with energy and
other resource limitations – but also with the *alternative means* (social actions)
by which these goals will be established and implemented by social institutions
and social leaders (Schnaiberg, 1980: ch. IX; Stretton, 1976).

The more sociologically and historically grounded view of the social response to the Schumacher-Lovins ideas is as follows. The AT–SEP ideology
(Morrison, 1978, ch. 11 in this volume) arises as a challenge (Gamson, 1975) to
existing ideologies that support the modern "treadmill of production" (Schnaiberg, 1980: ch. V). Social groups with strong *interests* in perpetuating the treadmill attempt to refute and/or suppress these competing ideas, since implementation of them would alter existing systems of privilege, which benefit these interest groups (capital and labor segments, for example). (Other groups resist as well
because of habitual lifestyle commitments). That is, one strategy is outright
resistance. A second strategy is *cooptation:* a superficial positive response to the
ideas and the movement carriers of these ideas. This includes invitations to
speak, appointments to boards of corporate or institutional directortes,
appointments to policy advisory committees to establish "balance," grant support for further research and even pilot plant development. A third strategy, of
course, is *capitulation:* firm acceptance of the new ideas, for a particular *time
period*. Following a dialectical model, eventually new resistance will arise from
newly competing interests – personal and/or structural.[2]

Only the capitulation model is consistent with the Schumacher-Lovins assumptions, and then only if capitalism is viewed as permanent (*cf.* Schnaiberg, 1981). It may be argued that a gradualist interpretation of Lovins and Schumacher suggests that capitulation is not a necessary condition for their agendas to be met. Such incrementalist interpretations assume that voluntary lifestyle changes (e.g., Hackett and Schwartz, 1980) or voluntary community development efforts (e.g., Pitts, 1981) will slowly displace the existing treadmill of production. However, this model assumes a supportive climate or "benign neglect" of such voluntaristic movements: increased resistance or increased cooptation of such voluntarism would still impede this gradualist or incrementalist transformation. And, in fact, American and European response to Schumacher and Lovins since 1973 or 1976 demonstrate primarily resistance, and secondarily cooptation (e.g., Primack, 1980; Pitts, 1981).

Resistance is so obvious a dominant mode of response that it does not warrant further discussion. Suffice it to say that the principal root for resistance appears to be a real and/or perceived *interest* in the personal-organizational benefits of a perpetuation of the inappropriate technology-hard energy path system of social production (e.g., Frahm and Buttel, 1980). This may include conventional social conflicts over the relations of production (e.g., Buttel and Larson, 1980), or more diffuse consumption differences (e.g., Hackett and Schwartz, 1980).

Whether such resistance will grow or shrink is a matter of conjecture (e.g., Schnaiberg, 1981). Part of the projection of resistance is dictated by the future trajectory of resources. But by far the larger part of future history will be dictated by the socioeconomic decision-making and commitments made by present and future economic and political elites (Buttel and Larson, 1980). Such decision-making is in part shaped by the success of current and future attempts to "deal with" the AT–SEP theorists as social problems claimants (Spector and Kitsuse, 1977). Recently, cooptation and capitulation have been used to supplement the politically more difficult route of outright resistance or repudiation (Mitchell, 1980). Since energy is a social and ecological resource (Schnaiberg, 1982a) that is central to both Schumacher's and Lovins' theories, it is instructive to examine recent energy "successes" claimed by the AT–SEP movement as a guide to patterns of cooptation and/or capitulation. Using recent American outcomes, the delineation of primarily cooptive strategies is outlined, and some conclusions about the general vulnerability of the models are induced from this specific case.

Success and Failure of AT-SEP in American Energy Policy-Making

Recent declines in growth rates of American energy consumption on both aggregate and per capita levels are frequently treated as indicators of institutionalization of the SEP program, following as they do on the heels of Amory Lovins articles, books, speeches, and guest appearances. Other indicators of successful incorporation of both Lovins' and Schumacher's ideas include the rise of "neighborhood technology" movements (Pitts, 1981) and local energy action (Ridgeway, 1979; Brunner, 1980), and "voluntary simplicity" (Hackett and Schwartz, 1980). The integration of earlier equity movements into local energy activity is of particular appeal to those who envision a new democratic participation technology such as Schumpeter and Lovins anticipate (e.g., Morris and Hess, 1975). Part of the ideological excitement is the integration of "equality" and "efficiency" (Okun, 1975) concerns in a single movement organization. This is contrary to much of American social history, in which such concerns have been located in different structures, which have not been integrated and indeed, have often been in considerable economic and political conflict (e.g., Friedland *et al.,* 1977; *cf.* Pitts, 1981). Elsewhere (Schnaiberg, 1982b, 1983b), I have suggested why this historical separation and/or goal conflict should caution us in projecting future successes for AT/SEP.

Following the timing arguments about Schumacher's and Lovins' works described above, the favorable climate for "soft energy" technologies and paths was very much in evidence in the 1970s. This, coupled with recent declines in energy consumption, seem almost prima facie evidence for movement success. However, a closer examination of the processes and outcomes underlying this assertion (Schnaiberg, 1981, 1982a) give the social analyst considerable pause and serious doubt as to whether this history indicates either capitulation or cooptation. Declines in U.S. energy consumption appear to be primarily the result of response to relatively unambiguous "signals" (Edney, 1980) about the *social* scarcity of energy in the U.S. The primary "signal" was the increase in the price of petroleum products (and other energy products), coupled with an earlier consciousness boost following the oil shortfalls induced by OPEC.

The political process by which this signal was reinforced was the decision to deregulate oil prices in the U.S., and allow the market to *efficiently* ration petroleum products. In Lindblom's (1977) terms this was in effect an abdication of "politics" to "markets"; the principle goal of this policy was to increase economic efficiency. This decision was supported by AT–SEP proponents, as *one* component of a social transition to an energy-conserving society, which was itself a way-station on the path to AT and/or SEP (*cf.* Primack, 1980). Even at this level, then, we could make a structural assessment that AT–SEP proponents had been coopted to support or acquiesce in having *efficiency* (Okun, 1975) as the primary socio-political *goal* of this policy, rather than *equality* (*cf.* Morrison, ch. 11 in this volume; Morrison and Lodwick, 1981; Mitchell, 1980). While movement proponents saw this as a tactical means of further *subverting* the current industrial treadmill, they nonetheless supported technical economic efficiency values. They might have opted instead for the broader concept of social efficiency, which is only loosely tied to economic efficiency (e.g., Mazur and Rosa, 1974). We can characterize this process as a *value-cooptation.*

Beyond this level of assessment, though, we need to examine how this non-AT/SEP goal was actually implemented. As any good economist (e.g., Thurow, 1980) could have predicted, it appears that the net effects of energy conservation have been regressive, because the *means* of achieving this goal have been efficient, but not at all equitable (*cf.* Morrison, ch. 11 in this volume). From the limited range of systematic data available (Schnaiberg, 1981), it would appear that a substantial share of energy conservation has derived from *capital replacement* in goods that use energy intensively. The social capacity to replace capital, in turn, has reflected both income and wealth. This means that (1) the more economically successful the organization or individual, the more readily they have adapted to the social scarcity of energy; and (2) the process was a technical adaptation to price considerations, devoid of both an ecological and a social calculus (*cf.* Hackett and Schwartz, 1980). If we combine these two processual elements, it seems clear that the U.S. implementation of energy conservation has reinforced the dominance of conventional *market* forces (e.g., Mitchell, 1980). It has also reduced consideration of social or ecological costs of permitting the market to operate (Schnaiberg, 1982a). Indeed, in an anthropomorphic sense, "the market has triumphed" (Baker, 1981), not the movement.

If we were to further dissect this recent experience to enhance our appreciation of the limitations of the models of Figure 1, we should first start with the left-hand side of the models. It is certainly true that this recent period has witnessed "new knowledge of system vulnerability". However, three questions immediately follow upon this assertion: (1) how accurate is this knowledge? (2) to whom has it been diffused? and for those groups to whom it has been diffused, (3) what have been the value-change consequences? In conventional terms, these questions can be translated into issues of (1) creation of social welfare and ecological intelligence (Schnaiberg, 1980: chs. VI, VII); (2) the dissemination of such social intelligence; and (3) the effects of such intelligence. For each of these domains, recent theorizing and empirical research indicates the importance of at least a conflict model of social change, if not a richer model of dialectical change.

Conflict, Cooptation and Resistance: Social Intelligence about American Energy Options

The three questions above relate to the creation, dissemination, and effects of social intelligence in the arena of energy decision-making in the U.S. Since space prevents a full treatment here of this issue, I will simply draw upon some major pieces of work in the broader area of environmental resource decision-making (e.g., Schnaiberg, 1977, 1980: chs. VI, VII; Meidinger and Schnaiberg, 1980), supplemented by specific extensions of this general model to energy decision-making in particular (e.g., Morrell, 1981; Burton, 1980).

Throughout this "social impact assessment" literature, there is generally an initial description of the new surge of interest in, and data on, issues of resource allocation in American production. That is, there is at the outset a sense of *capitulation* to the AT-SEP perspective, at least at the level of social intelligence, i.e., in research, though not necessarily research and development [R and D]. However, virtually every scholarly account immediately cautions us about (1) the limits of social resources for necessary research, and (2) the subsequent *ineffectiveness* of existing research findings in influencing decision-making by political and corporate elites (e.g., Morrell, 1981; Burton, 1980). In general, this is attributed to existing *interests* of these elites, and their resistance to AT-SEP evaluations because of the anticipated "negative" consequences for their affiliated production organizations, agencies, and social networks (Buttel and Larson, 1980).

The general picture that emerges is that the social creation of such information is less a technical than a *political* problem (*cf.* Friedland *et al.*, 1977). Methodological problems abound in this area, to be sure (e.g., Meidinger and Schnaiberg, 1980). But they are in principle resolvable with the application of sustained and adequate research funding, and the social desire to make use of the resulting data for resource decision-making. In addition, the problems of communicating the limited data that have been created by scientists and engineers are bemoaned by both social scientists and social movement participants (e.g., Schnaiberg, 1975, 1980: ch.VII). Again, the dominant report in the literature is that competing interests produce individual and organizational resistance, not attentiveness to challenge ideologies about the benefits of soft vs. hard energy paths, or appropriate vs. inappropriate technology. And this is true despite the institutionalization of environmental protection goals in the state (Buttel and Larson, 1980).

At these two levels, at least, we find strong evidence of cooptation and resistance in the actual patterns of gathering of data and diffusion of information. The only evidence that is strongly to the contrary is produced by observers who confuse legislative *intent* with actual policy *implementation* (e.g., *cf.* Culhane, 1974; Friesema and Culhane, 1976; Fairfax, 1978; Lowi, 1979). While there are undoubtedly important individuals and groups whose attitudes on resource allocation in general and energy in particular have been changed by research and communication through professional and media routes (e.g., Mitchell, 1978; Council on Environmental Quality, 1980), this does not appear to be a dominant outcome. That is, some *cooperation* with this new intelligence gathering has unquestionably arisen in the past decade—but this is far from capitulation, and given recent American political trends, it is unclear whether we are moving to less rather than more cooperation (e.g, Dickson, 1981; Wolin, 1981; Mitchell, 1980).

Unlike the models of Schumacher and Lovins, moreover, the absence and/or diminution of cooperation is not a matter of social *ignorance*. Rather, there is competing intelligence from interests that would be threatened by the production and application of new knowledge. Moreover, for *tactical* political reasons *some* support of the AT–SEP ideas and research may be necessary to minimize transaction costs of doing business or governance. But this should not be confused with acceptance of these ideas or findings: they are instances of cooptation as the preferred path, in contrast to outright resistance (e.g., Fairfax, 1978). Had they indeed been the first stages of a growth in acceptability of ideas, there would not have been such a strong shift in opposition to such ideas with the emergence of the "reindustrialization" ideology of the Reagan regime, for example (Schnaiberg, 1981). The fact that there are parallel shifts in American political ideology towards a strengthening of traditional capital- and energy-intensive production does not exempt this U.S. trend from consideration of resistance patterns. Indeed, it strengthens the case that corporate and political resistance is far more widespread than AT–SEP movement proponents willingly acknowledge (Primack, 1980). Social scientists have colluded in this misrepresentation of movement accomplishments by focusing primarily on trends in protest movement inputs rather than on their actual achievements and/or *counter*movement actions (*cf.,* Goldstone, 1980; Walsh, 1981; Barkan, 1979; Hornback, 1974; Pitts, 1981).

To briefly complete this picture of cooptation and/or resistance, it is important to stress that the *effectiveness* of new data and perspectives on resources is even more restricted than the Schumacher-Lovins model would suggest. If we measure this effectiveness by actual changes in production policy or organization, and/or concrete commitments to future changes (near- or longer-term), we find extremely limited effectiveness (e.g., Morrell, 1981; Burton, 1980) in the case of energy policy – the hallmark of the AT–SEP program of social change. Here it is even more clearly the actual or perceived interest in competing modes of production (Burton, 1980: 298–302) that is at the core of resistance. Interestingly, the sole measure of "success" of the AT–SEP "program" or model is the reduction in American energy consumption, where the outcomes have been inegalitarian. If this is "success", then one might wonder whether failure would have been preferable for the long-term welfare of American working and poverty-classes (*cf.* Morrison, ch. 11 in this volume) – or even for the broad middle class (Schnaiberg, 1981, 1982a; Primack, 1980). The breadth and depth of the *productive changes* proposed by AT–SEP movements is enormous, unlike the range of most protest movements (e.g., Gamson, 1975, 1980; Ash, 1972; Goldstone, 1980). Nothing less than a total restructuring of production and consumption is being advocated.[3] But at the same time, societal survival becomes more and more the responsibility of AT–SEP movement advocates, if their ideas are taken more seriously. This leads to further threats of cooptation and resistance, even when initial success occurs, since: "If one accepts responsibility for the maintenance of a general welfare, including that of one's opponents, one is on the path to corruption and impotence. . . It is easy [though] to maintain one's integrity when one's words and actions are ineffective" (Ravetz, 1971:428).

A simplified illustration of this principle is the basis of the next section, in which the technical and welfare requirements of the "transition to appropriate technology" in energy are elucidated.

The Dilemma of the Transitional Period: An Illustration of Energy Policy Conflicts

At a theoretical and broad empirical level, it is relatively easy to foresee ongoing conflicts because of competing interests and their socioeconomic and political representation. Yet a forceful statement of AT–SEP goals, arising from a powerful depiction of system vulnerability, often distracts social scientists' attention away from enduring conflicts (e.g., Dickson, 1974, 1981; Cooper, 1972, 1973). One way of grounding our sense of future trajectories of conflict is to examine a projection/scenario when there *is* widespread acceptance of system vulnerability and thus greater acknowledgement of the need for new technologies to adapt to social scarcity of energy.

Drawing upon a recent article by Rossin (1980), I will outline likely conflicts even where HET (Hard Energy Technology) proponents apparently capitulate on the need for soft energy technologies. Rossin, a nuclear engineer employed by a major midwestern utility, faithfully represents the interests and some of the *emerging* strategies of the HEP utility industry. His article reflects the mixture of resistance, cooptation, and cooperation utilities have used in response to the AT–SEP movement.

Rossin, as if scripted by Ravetz, argues that SEP proponents are taking on a huge social burden if their proposals are followed. For the continuity of social welfare, he convincingly pushes for a triple role of HET: (1) as a *transitional* source of socially-necessary energy to maintain the social system; (2) as a form of social *vulnerability* insurance when there are problems with renewable energy (wind, solar) under SEP technologies; and (3) as a lower social cost *alternative* to renewable decentralized energy in many locales and applications. Ironically, in many ways he has taken the arguments of Lovins and others, and found ways of redirecting these positions to *favor* hard energy paths in particular times, areas, and applications. However the details are described, though, the essential argument of Rossin is for *perpetual coexistence* of hard and soft energy paths. This is something which both Schumacher and Lovins grudgingly permit, though do not necessarily encourage.

Perpetual coexistence of high-capital and lower-capital energy sources appears socially sensible. But the price for this reduction of social risk is perpetual conflict between SEP and HEP interest groups. Capitalist and even socialist economies (Schumpeter, 1947; Goldman, 1972) strongly dictate this. Investors with an interest in existing high-capital centralized energy plants must be making a profit (fiscal or social) in order for such plants to be maintained with depreciation covered. Under these conditions, it is everywhere and at all times in these investors' (and the corporate managers') interests to *expand* this profit-making venture, so long as profitability here exceeds the profitability in other ventures. If this is so, then the economic interests of utility investors stand in direct competition with the social goals and means of the SEP proponents, as with any other form of economic competition (e.g, Tanzer, 1971). That is the essential logic of profitability, with socialist equivalences.

The "simple" way out of this apparent dilemma of perpetual conflict is to assure a rate of return of HET no higher than that on any other investment. This is the historical rationale for allowing public utility commissions to control utility monopolies and their rate structures. But by this argument, the rate can fall *no lower* than that for other investments, either. Thus, HET will still remain exactly "competitive" with all other investment opportunities. If it falls too low, then plants will decay with depreciation. If these plants are needed, one can argue, the state will become an investor. But the state apparatus – as the complex history of the Tennessee Valley Authority clearly indicates – often behaves identically with large private investors, because of common need for capital rationing (Schumpeter, 1947.) So state control merely transfers the competition between HET and SET to a movement-state conflict, rather than a movement-private investor conflict (e.g., Goldman, 1972; Buttel and Larson, 1980).

Note that this analysis grows out of a model of apparent capitulation of HET proponents to the SEP agenda, though Rossin's analysis excludes value and consumption changes. The fact that his persuasive argument implies a rather deeper and more persistent level of future conflict only underscores the complexity of the transition. It also points to the unspecified nature of the "transformed" productive system (Schnaiberg, 1982b). More importantly, it describes an enduring base for conflict – the economic interests and political influence of powerful economic interest groups – itself a product of social and technological history, with its outcomes of dominant and subordinate groups. Past generation of energy and its allocation through markets (and politics) has produced highly stratified societies, in terms of the capacity to mobilize, allocate, and protect capital. As well, it has produced, as Harvey (1973) stresses, some immobility of fixed capital in the form of operating equipment. Investors in this capital often have both the *capacity* and the *need* to resist technological changes that will undermine their profits and hence reduce their economic freedom to invest and control investment. This element of immobility and competition is virtually absent from the AT–SEP scenarios, and its absence dictates that this model is like a vehicle without an engine – it has form but no capacity for movement (Buttel and Larson, 1980).

Conclusions: Can These Models be Saved? Should They Be?

Three quite different conclusions can be drawn from the above assessment. First, we can assert that the AT–SEP models (predictive or prescriptive) are so sociologically naive that they are unsalvageable as social science – and likely of limited durability as social movement ideology as well. Our second option is to broaden these models to take account of social conflict and social interests, and to incorporate greater predictions about the nature and intensity of conflict as the social movements adhering to these models' struggle for true cooperation and capitulation, as against resistance and cooptation (*cf.* Morrison and Lodwick, 1981).

But the AT/SEP models represent both normative positions (we *should* follow AT–SEP rather than the current path) as well as sets of predictive statements (if we follow AT–SEP, *then* these following consequences will occur). This suggests two roles for social scientists *as scientists*. First, we need to focus more clearly on the normative assumptions of the models, and separate them from predictive statements. Second, we should apply all our knowledge of the historical and contemporary functioning of the social system to *evaluate* the validity of these social claims (Spector and Kitsuse, 1977; Morrison and Lodwick, 1981).

Modestly, we must acknowledge that we do not have the scientific capacity presupposed in the model of "social impact assessment" required under the National Environmental Policy Act (e.g., Finsterbusch and Wolf, 1977; *cf.* Meidinger and Schnaiberg, 1980). However, many of the changes proposed by AT–SEP are so sweeping that we could probably evaluate *these* with more confidence. First, though, we need to be far more realistic about distinguishing changes in the *forces* of production (physical technology) from those in the *relations* of production (social class structure) in these models of change (*cf.* Etzkowitz, 1981). Second, we should attempt to project which types of changes are likeliest to occur (e.g., changes in some physical technology of energy production), and which are least likely to occur (e.g, equalization of access to energy), given the existing class structure and likely trajectory of political conflict (Buttel and Larson, 1980).

Having done this, social scientists could be in a position of evaluating the socially probable outcomes of the AT–SEP movement. This will take account of the contending forces (e.g., Tucker, 1981; Rosencranz, 1981; Alexander, 1981) favoring either an economic synthesis or a managed scarcity approach to resource scarcity (Schnaiberg, 1975, 1980: ch. IX). It will also take account of the conflict and hardship imposed by rationing of scarcity that these opposing claims entail (e.g., Schnaiberg, 1975; Primack, 1980; Pitts, 1981). From this set of predictive statements, a crude form of social impact assessment of AT–SEP *movements,* we can then allow informal political processes to continue.

In short, we ought not to become movement advocates or opponents, in our role as social scientists. We ought to raise political consciousness about implications for the future functioning of both ecosystems and the social welfare system.

Notes

1. On the basic arguments of appropriate technology and soft energy paths, see

Schumacher (1973) and Lovins (1976, 1977). For a review of the American energy policy-making see Congressional Quarterly, Incorporated (1981). Some critical evaluation of the Lovins approach is in Morrison and Lodwick (1981), and the references cited therein. For Schumacher's work, which was originally oriented to Third World development, see Morrison (1978) and the critical evaluations of David Dickson (1974: ch. 6) and Charles Cooper (1972, 1973). While Cooper is focusing on *international* "technology transfer" politics, his analysis of class structure and class interest effects applies to U.S. *domestic* technology transfers as illustrated by the empirical analysis of Pitts (1981) and Mitchell (1980), and the theoretical critique of Schnaiberg (1982a) and Buttel and Larson (1980).

2. The recent debate between Goldstone (1980) and Gamson (1980) indicates the difficulty of empirically judging whether cooptation or capitulation can be readily distinguished in the "success" of a social movement (*cf.* Gamson, 1975). Goldstone's central thesis is that American movements with goals that exclude *displacement* of elites are more "successful" than those seeking to displace elites. Neither Schumacher nor Lovins indicate the *degree* of displacement of elites necessary for AT or SEP. Schumpeter (1947: ch. XVIII) gives a very thoughtful review of displacement needs for socialist transformation, which ought to be reviewed in this AT–SEP proposal for social change as well (*cf.* Stretton, 1976). Seiler and Summers (1979) illustrate the *empirical* problems of distinguishing cooptation from capitulation.

3. In Gamson's (1975) and Goldstone's (1980) terms, it is hard to imagine this transformation occuring without *displacement* of current economic elites temporarily, if not permanently. This is the root of conflict arising from competing interests.

References

Alexander, Tom. 1981. "A simpler path to a cleaner environment." *Fortune* 103(May 4):234ff.

Ash, Roberta. 1972. *Social Movements in America.* Chicago: Markham.

Baker, Wayne. 1981. Markets as Networks: A Multimethod Study of Trading Networks in a Securities Market. Unpublished doctoral dissertation, Dept. of Sociology, Northwestern University, Evanston, IL.

Barkan, Steven E. 1979. "Strategic, tactical, and organizational dilemmas of the protest movement against nuclear power." *Social Problems* 27(October):19–37.

Brunner, Ronald D. 1980. "Decentralized energy policies." *Public Policy* 28(Winter): 71–91.

Buttel, Frederick H. and O.W. Larson, III. 1980. "Whither environmentalism? The future political path of the environmental movement." *Natural Resources Journal* 20(April):323–344.

Burton, Dudley J. 1980. *The Governance of Energy: Problems, Prospects and Underlying Issues.* New York: Praeger-Cambria.

Commoner, Barry. 1977. *The Poverty of Power: Energy and the Economic Crisis.* New York: Bantam.

Congressional Quarterly, Inc. 1981. *Energy Policy.* Second edition. Washington, D.C.: Congressional Quarterly, Incorporated.

Cooper, Charles. 1972. "Science, technology and production in the underdeveloped countries: an introduction." *Journal of Development Studies* 9(October):1-37.

————. 1973. "Choice of technique and technological change as problems in political economy." *International Social Science Journal* XXV (3):293-304.

Council on Environmental Quality *et al.* 1980. *Public Opinion on Environmental Issues: Results of a National Public Opinion Survey.* Washington, D.C.: Government Printing Office.

Culhane, Paul J. 1974. "Federal agency organizational change in response to environmentalism." *Humboldt Journal of Social Relations* 2(Fall/Winter):31-44.

Dickson, David. 1974. *Alternative Technology and the Politics of Technical Change.* Glasgow: Fontana/Collins.

————. 1981. "Limiting democracy: technocrats and the liberal state." *Democracy* 1(January):61-79.

Edney, Julian J. 1980. "The commons problem: alternative perspectives." *American Psychologist* 35(February):131-150.

Enloe, Cynthia H. 1975. *The Politics of Pollution in a Comparative Perspective: Ecology and Power in Four Nations.* New York: David McKay.

Etzkowitz, Henry. 1981. "Hyperdeterminisme: ideologies de l'energie et de l'environnement aux Etats-Unis." *Sociologie et Societes* XIII (1):63-80.

Fairfax, Sally K. 1978. "A disaster in the environmental movement." *Science* 199(17 February):743-748.

Finsterbusch, Kurt and C.P. Wolf, editors. 1977. *Methodology of Social Impact Assessment.* Stroudsberg, PA: Dowden, Hutchinson, and Ross.

Frahm, Annette, and Frederick H. Buttel. 1980. "Appropriate technology: current debate and future possibilities." Unpublished manuscript, Dept. of Rural Sociology, Cornell University, Ithaca, NY.

Friedland, Roger; F.F. Piven; and Robert R. Alford. 1977. "Political conflict, urban structure, and the fiscal crisis." *International Journal of Urban and Regional Research* 1(October):447-471.

Friesema, H. Paul, and P.J. Culhane. 1976. "Social impacts, politics, and the environmental impact statement process." *Natural Resources Journal* 16(April):339-356.

Gamson, William A. 1975. *The Strategy of Social Protest.* Homewood, IL: Dorsey Press.

————. 1980. "Understanding the careers of challenging groups: a commentary on Goldstone." *American Journal of Sociology* 85(March):1043-1060.

Goldman, Marshall I. 1972. *The Spoils of Progress: Environmental Pollution in the Soviet Union.* Cambridge, MA: M.I.T. Press.

Goldstone, Jack A. 1980. "The weakness of organization: a new look at Gamson's *The Strategy of Social Protest." American Journal of Sociology* 85(March):1017-1042.

Hackett, Bruce, and S. Schwartz. 1980. "Energy conservation and rural alternative lifestyles." *Social Problems* 28(December):165-178.

Harvey, David. 1973. *Social Justice and the City.* Baltimore, MD: Johns Hopkins University Press.

Hornback, Kenneth E. 1974. Orbits of Opinion: The Role of Age in the Environmental Movement's Attentive Public, 1968-1972. Unpublished doctoral dissertation, Dept. of Sociology, Michigan State University, East Lansing, MI.

Inglehart, Ronald. 1977. *The Silent Revolution: Changing Values and Political Styles among Western Publics.* Princeton, NJ: Princeton University Press.

Leonard-Barton, Dorothy. 1980. "The role of interpersonal communications networks in the diffusion of energy-conserving practices and technologies." Paper presented at International Conference of Consumer Behavior and Energy Use, Banff, Alberta, Canada, September.

_____, and E.M. Rogers. 1981. "Horizontal diffusion of innovations: an alternative paradigm to the classical diffusion model." Working paper 1212-81, Sloan School of Management, Massachusetts Institute of Technology.

Lindblom, Charles F. 1977. *Politics and Markets: The World's Political-Economic Systems.* New York: Basic Books.

Lovins, Amory. 1976. "Energy strategy: the road not taken?" *Foreign Affairs* 55(October):65-96.

_____. 1977. *Soft Energy Paths: Toward a Durable Peace.* New York: Harper Colophon.

Lowi, Theodore J. 1979. *The End of Liberalism.* Second edition. New York: W.W. Norton.

Marsh, Alan. 1975. "The silent revolution, value priorities, and the quality of life in Britain." *American Political Science Review* 69(March):21-30.

Mazur, Allan, and E. Rosa. 1974. "Energy and life-style." *Science* 186(15 November): 607-610.

Meidinger, Errol, and A. Schnaiberg. 1980. "Social impact assessment as evaluation research: claimants and claims." *Evaluation Review* 4(August):507-535.

Mitchell, Robert C. 1978. "Environment: an enduring concern." *Resources* 57(January-March):1ff.

_____. 1980. "How 'soft', 'deep', or 'left?' Present constituencies in the environmental movement for certain world views." *Natural Resources Journal* 20(April):345-358.

Morrell, David. 1981. "Energy conservation and public policy: If it's such a good idea, why don't we do more of it?" *Journal of Social Issues* 37(Spring):8-30.

Morris, David, and K. Hess. 1975. *Neighborhood Power: Returning Political and Economic Power to Community Life.* Boston: Beacon Press.

Morrison, Denton E. 1978. "Energy, appropriate technology, and international interdependence." Paper presented at annual meetings of the Society for the Study of Social Problems, San Francisco, September.

_____. 1980. "The soft cutting edge of environmentalism: why and how the appropriate technology notion is changing the movement." *Natural Resources Journal* 20 (April):275-298.

Morrison, Denton E., and D.G. Lodwick. 1981. "The social impacts of soft and hard energy systems: the Lovins's claim as a social science challenge." *Annual Review of Energy* 6:357-378.

Okun, Arthur. 1975. *Equality and Efficiency: The Big Tradeoff.* Washington, D.C.: The Brookings Institution.

Pimentel, David; L.E. Hurd; A.C. Bellotti: M.J. Forster; I.N. Oka; O.D. Sholes; and R.J. Whitman. 1973. "Food production and the energy crisis." *Science* 182(2 November):443-449.

Pitts, James P. 1981. "Urban neighborhoods and appropriate technology: a case study in technology transfer." Paper presented at annual meetings of the American Sociological Association, Toronto, Canada, August.

Primack, Phil. 1980. "Soft energy and hard times." *Working Papers for a New Society* VII (July/August):15–23.

Ravetz, Jerome R. 1971. *Scientific Knowledge and its Social Problems.* Oxford: Clarendon Press.

Ridgeway, James. 1979. *Energy-Efficient Community Planning.* Emmaus, PA: J.G. Press. Inc.

Rosencranz, Armin. 1981. "Economic approaches to air pollution control." *Environment* 23(October):25–30.

Rossin, A. David. 1980. "The soft energy path: Where does it really lead?" *The Futurist* 14(June):57–63.

Schnaiberg, Allan. 1975. "Social syntheses of the societal-environmental dialectic: the role of distributional impacts." *Social Science Quarterly* 56(June):5–20.

_____. 1977. "Obstacles to environmental research by scientists and technologists: a social structural analysis." *Social Problems* 24(June)500–520.

_____. 1980. *The Environment: From Surplus to Scarcity.* New York: Oxford University Press.

_____. 1981. "Energy, equity, and austerity: some political impacts of energy rationing by price." Paper presented at annual meetings of the Society for the Study of Social Problems, Toronto, Canada, August.

_____. 1982a. "Energy conservation in the U.S.: a pyrrhic social victory?" Unpublished manuscript, Department of Sociology, Northwestern University, Evanston, IL., December.

_____. 1982b. "Did you ever meet a payroll? Contradictions in the structure of the appropriate technology movement." *Humboldt Journal of Social Relations* 2 (Spring/Summer):38–62.

_____. 1983a. "Saving the environment: from whom, for whom, and by whom?" In Nicholas Watts and P. Knoepfel (eds.), *Environmental Politics and Policies: An International Perspective.* Frankfurt and New York: Campus Publishers, forthcoming.

_____. 1983b. "Redistributive goals vs. distributive politics: social equity limits in environmental and appropriate technology movements." *Sociology Inquiry,* forthcoming

Schumacher, E.F. 1973. *Small is Beautiful: Economics As If People Mattered.* New York: Harper & Row.

Schumpeter, Joseph A. 1947. *Capitalism, Socialism, and Democracy.* Third edition. New York: Harper and Brothers.

Seiler, Lauren H., and G.F. Summers. 1979. "Corporate involvement in community affairs." *The Sociological Quarterly* 20(Summer):375–386.

Spector, Malcolm, and J.I. Kitsuse. 1977. *Constructing Social Problems.* Menlo Park, CA: Cummings Publishing Company.

Stretton, Hugh. 1976. *Capitalism, Socialism and the Environment.* Cambridge: Cambridge University Press.

Tanzer, Michael. 1971. *The Sick Society: An Economic Evaluation.* Chicago: Holt, Rinehart, and Winston.

Thurow, Lester C. 1980. *The Zero-Sum Society: Distribution and the Possibilities for Economic Change.* New York: Basic Books.

Tucker, William. 1981. "Marketing pollution." *Harpers* 262(May):31–38.

Walsh, Edward J. 1981. "Resource mobilization and citizen protest in communities around Three Mile Island." *Social Problems* 29(October):1-21.

Watts, Nicholas, and G. Wandesforde-Smith. 1980. "Postmaterial values and environmental policy change." *Policy Studies Journal* 9(Winter):346-358.

Wolin, Sheldon S. 1981. "The people's two bodies." *Democracy* 1(January):9-24.

Chapter 13

Economic Effects of Technology in Agriculture in Less Developed Countries

William C. Thiesenhusen

Given the phenomenal successes of United States agriculture, it may seem odd to argue that some technological and institutional "package" other than the one we used might be more appropriate for less developed countries (LDCs). This, however, is what I wish to do in this paper. I will: (1) define what seems to constitute "appropriate technology;" (2) point to the successes of the food system in the U.S. with reference to the appropriate technology it utilized (while acknowledging its weaknesses);[1] (3) show that highly skewed distribution of income and wealth and high incidence of unemployment makes the institutional system in most LDCs much different from that which prevailed in the U.S. while it was industrializing; (4) argue that the problems LDCs confront are at least as much social as technological; and (5) suggest that U.S.-type institutional structures and technologies are usually not appropriate – at least not without much adaptation – for agricultural development in emerging nations.

What is Appropriate Technology?

The technological package that is used to develop agriculture in LDCs must have at least two social requirements. First, the technology itself must be appropriate

The author has benefited from comments on earlier drafts from his colleagues Dr. Jane Knowles, Professor Peter Dorner, Professor Marion Brown, and Ms. Jane Dennis-Collins. They bear no responsibility for error or misinterpretation which must be ascribed solely to the author.

to the factor proportions of the country in question. Secondly, the institutions through which it is delivered must be adequate to the task. If one or the other is distorted, production may drop and/or any of the other attributes which an optimal rural development program should have may fall short of expectations. If technology is inappropriate it will have strong and probably undesirable side effects (viz., unemployment may accompany the production of a marketable surplus) which detract from the major reason the technology was introduced in the first place. A major role of institutions in the adoption process is that they may hinder or facilitate the receipt of technology by one group or another within a society.

Of course, not all technology is highly sensitive to its milieu. Some non-economic kinds are almost as useful in one ambience as another. Take, for example, the measles vaccine. Whether the economy is advanced or lagging, whether the recipient is rich or poor, whether the distribution of income is egalitarian or skewed, the vaccine will be effective provided it gets to its recipient and it usually does. Almost nothing prevents its diffusion because, as the work of WHO has shown, the diffusion and capital costs are relatively low.

In agriculture, the process of development in the United States called forth a kind of technology that responded with a fair amount of accuracy to factor scarcity. Since labor was the factor that was most often at a premium, labor-saving machinery came to be profitable. When the frontier closed and war demanded that more be produced, land-saving technology in the form of hybrid seeds and fertilizers were added to the arsenal of agricultural technology. The oil scarcity of the 1970s set off a search for ways to save it or stretch it as far as possible and we do not as yet know how this creative flux will work itself out.

When LDCs got around to beginning the development process a range of technology from the developed countries (DCs) was available to them. The temptation for them to choose as much as possible was understandable. The combination had produced spectacular results in developed countries and was endowed in the minds of many experts with almost mystical qualities. Furthermore, sometimes it could be gotten gratis through aid programs. Or private industry might have been able to acquire it in a highly subsidized manner.

The problem is that the institutional pattern is different among the LDCs and the factor scarcity pattern is likewise distinct. So in some cases when the labor factor is plentiful and management decides to use machinery, for example, unemployment may be the result. This probably will not happen as quickly in LDCs in which the owner-operator's family is the labor force. The owner-operator probably will not fire his family (though it may permit a large amount of underemployment). In this case the family farm institutional pattern can be thought of as having raised the adoption costs for inappropriate technology. Having jobless family members is too high a price for an owner-operator to pay for utilizing labor-saving technology.

Using technology to substitute for labor even when ample labor supplies exist came to the attention of development planners as one example of an unfortunate side effect of technology borrowing by LDCs. It fostered some anti-technology arguments for a time and then there was a call for technology of a different sort. E.F. Schumacher utilized the term "intermediate" rather than the currently used "appropriate" technology to describe it. He did not realize that in so doing he would set off a rather heated semantic debate (Howes, 1979). Before he wrote *Small Is Beautiful,* Schumacher elaborated some criteria for appropriate technology which are still applicable (1971:85-94):

1. Jobs have to be created in areas where people are living now — not primarily in metropolitan areas into which they tend to migrate.
2. These must be cheap enough so they can be created in large numbers without requiring an unattainable level of savings and inputs.
3. The production methods employed must be relatively simple so demands for high skills are minimized — not only in the production process itself but also in matters of organization, raw material supply, financing, marketing, etc.
4. Production should be largely from local materials.

These can be met, he claimed, only if there is a regional approach to development and if there is a conscious effort to develop intermediate technology.

Schumacher was a first entrant into a long and acrimonious debate from which I shall have time to merely sample. In 1977, Jedlicka (1977:10) broadly defines appropriate technology as "one that effectively utilizes the manpower, resources, and environmental and institutional realities . . . in a given country." Swanberg states the issue clearly, but the concept becomes more complex.

Improved technologies may come in a variety of forms, but may require an increase in the use of some resources. If the farmers are limited in their access to these resources, a constraint arises. It follows, then, that only when a new technology package faces no constraints can a technology properly be labeled as "appropriate." However, when one considers the resource requirements of an improved technology and observes a farmer's corresponding limitations, more than just "technical" inputs must be analyzed. Not only does one have to look at the bio-physical input combinations, including response to fertilizer, . . . [but at] socio-economic aspects such as labor profiles, liquid asset levels, arable land, market incentives, and product and input prices. . . . In addition, the organizations which facilitate the use of the farmer's own or borrowed credit must be reviewed. . . . Lastly, the concept of risk must be dealt with adequately. Final harmony between requirements and limitations must be attained if one expects the improved technology to be adopted by small scale farmers, before technologies can claim the classification "appropriate." (Swanberg, 1980: 1-2)

Swanberg's points were expressed succinctly by Hulse, who notes (1982: 1294): "Scientific principles are universally transferable; many technologies are not. Technologies based on biological principles, whether they relate to the cultivation of edible plants or to the transformation, preservation, and distribution of plant and animal products, are intensely influenced by their surrounding physical, social, and economic environments."

Chambers suggests that arriving at decisions on appropriate technology involves a thought process consisting of three steps. First, the characteristics of a "desirable rural future" are specified. This involves assigning certain weights to values and, hence, will always be open to some argument. Second, the anticipated relative endowments of the factors of production are assessed for a particular rural environment. Third, the characteristics of the technology which links the resource endowments to the desired endowment must be specified. If the technology does not exist, the challenge is to create it. Chambers lists five desirable characteristics for "future rural environments" (1977:28): (l) stability of the ecological system; (2) the productivity of whatever resources are scarce should be high rather than low; (3) more rather then fewer livelihoods should be generated from the system; (4) incomes and food supplies should be generated over the entire year and not just part of it; and (5) the benefits derived from resources should be distributed more rather than less equitably among the population.

It seems obvious that the state of the arts on appropriate technology is not yet very far advanced. It is certain that without appropriate technology countries tend to waste resources (like energy) and often dislocate people. It is clear that whatever technology is used in poor countries, it must maximize returns to the scarce factor. It is also obvious that we can arrive at a combination of appropriate technologies only by analyzing the existing institutional patterns of that society through which the technology must be filtered. It would seem worthwhile at this point to attempt an analysis of the technology and farm institutions utilized as agriculture in the United States developed in an effort to point up how a similar package tends to fail in many LDCs.

The U.S. Model: A Glance

We have become accustomed to even "family farming" being conducted on a large scale in the United States, with an enormous amount of capital necessary to be competitive.[2] Our technology is designed to serve big units. A generation ago, two-row corn planters were the mode; now many Wisconsin farmers plant fourteen rows at a time. The standard two-bottom moldboard has been replaced by machines that till a swath four times as wide. Weeds were pulled or chopped by hand. Now they are eliminated by herbicides, dispersed from large spray rigs. Small grains were shocked and, when they were dry, perspiring neighborhood crews put them through a stationary threshing machine attached by belt to the power drive of a tractor (earlier the steam engine provided the power). Now on a trip through the nation's breadbasket at harvest time one sees enormous self-propelled combines operated from an air-conditioned cab which may also be equipped with a tape deck. One farm machinery manufacturer proclaims that soon the driver may be eliminated, replaced by laser beams which will guide machines along flawlessly straight rows. The farmer will not be in the field, but at the controls of a computerized system that manages all aspects of an increasingly complex and ever more cerebral operation.

Most of the dramatic changes that have brought us to the brink of auto-mated agriculture have occurred in just 25 years. As industry grew and as fairly inexpensive education was available to farm youth, they left rural areas. And as they did, technology replaced physical exertion with motor power, gasoline, and chemicals. This technology sometimes took on a life of its own, in some cases pushing whole families—even those who wanted to stay—off the land and pre-venting some newcomers from entering agriculture. As land and necessary equipment came to be expensive, the opportunity to buy a farm and machinery slipped out of the reach of individuals with no family ties to agriculture. And making a living for those already on a farm meant keeping up with the "agri-cultural treadmill": early adopters bought newly developed machines, which lowered their cost per unit of output and increased their profits for several years; as more and more farmers purchased equipment, commodity prices dropped so that the remaining nonadopters were forced out of farming. They sold or rented their property, and the process of "technology introduction/land concentration" continued.

Some of those who left—many, perhaps—remained in farm-related indus-tries and professions. But only about 3 percent of the U.S. population lived on farms in 1978, and 2.7 percent in 1980. Food and fiber industries, however, are still the nation's biggest employers: 14 to 17 million people work in some phase of agriculture, from the field to retail sales. Most of these sales are in the manufac-turing sector, for as soon as production methods modernize, agriculture sloughs the sub-sector off—or manufacturing gobbles it up. Therefore, butter and cheese are no longer produced on the farm, for example. The most inefficient processes —those attendant upon growing the raw materials—remain in agriculture. Twenty percent of the employment in manufacturing in 1980 was agriculture-related.

Several other facts are important to an understanding of U.S. agriculture: agriculture is the country's largest industry, with assets in 1981 estimated at over $980 billion; since 1971, U.S. agricultural exports have expanded nearly sixfold, to a record $41 billion in 1980, making a net contribution of $24 billion to the U.S. balance of payments; the country supplies about half of the grain that moves in world trade and three-fourths of the soybeans. In 1981 the United States had a record 327 million metric ton grain crop, thanks to good weather and the most modern farming practices in the world.

Whether all of this can be sustained in the face of falling profit margins at the farm level remains to be seen. Indeed, the institutional patterns that have developed in agriculture have ensured that U.S. farmers since the 1940s have captured very few of the benefits of better technology in the form of bigger profits. Because there are so many farmers producing a relatively homogeneous product, they have no control over output price. And big production with relatively inelastic demand means quickly dropping real prices. Inputs, on the other hand, are produced under monopolistic competition. Today's operating costs are made up of three items that have risen especially dramatically of late: real estate taxes, energy, and interest payments on borrowed capital. This means that total real costs (discounting for those which new technology brings) have an upward bias; farm prices a downward bias; and farmer profits are represented by the space in between the two jaws of the closing vise. Turning the internal terms of trade against agriculture may have disastrous effects on farmers but be a boon to consumers. The fact that real food prices in fact rise is due to what happens after the raw materials leave the farm gate and enter into channels in which oligopoly prevails. The situation is one in which almost everyone is helped but the farmer. Big gainers from the workings of the treadmill are all consumers (their incomes are rising faster than food prices), and processors, wholesalers, and retailers— almost everyone beyond the farm gate. Farmers are constantly trying to keep profits up by increasing the quantity produced. Government subsidization of some commodities is one way of redressing the "subsidy" through the market place which farmers provide to the urban consumers.

The LDCs' Dependence on the United States

World dependence on this system makes importing countries vulnerable to the exigencies of U.S. agriculture. The United States and three other countries account for 90 percent of the world's farm exports; the U.S. provides about 70 percent of all the food aid of the world, though total food aid is relatively insignificant since the vast majority of agricultural products move in commercial channels.

The picture in less developed countries is quite different. Even if presently low levels of per capita consumption of cereals are to be maintained—given present distribution of wealth and income—output will have to increase by something like 30 million tons a year, which amounts to two-thirds of the LDC average imports during the years 1970-75. By the year 2000, with a projected world population of about 6.2 billion, the grain shortfall will have increased by 70 percent (Press, 1978:739). Even now there is the realization that though world output has more than doubled since 1950, the food machine is slowing down. Most gains came in the 1950s and 1960s, and "slow growth" occurred in the 1970s.

It seems clear that LDCs must be gearing up to produce more of their own food supplies.

From what I have said thus far, one could easily conclude that the problems in the farm sector are confined to technological problems of producing more food. Were this the case, the agricultural research agenda proposed by President Carter's science and technology advisor several years ago could remain unamended. To be sure, there is a pressing and undisputed need for research which will make countries less dependent on the United States and more self-sufficient. This means work that will make the humid tropics and arid regions more productive. Since most R&D has taken place (and is still located) in developed countries (98 percent by some estimates), most of the vast strides in the application of science to farming to date have been in temperate farming. Even the "Green Revolution," fathered in large part by International Agricultural Research Centers in Mexico and the Philippines, is firmly rooted in Western funding sources, and, on balance, the largest gains in output knowledge have come in temperate crops. The same can be said for the recent revolutionary developments in agricultural genetics, commonly referred to as gene splicing.

But this omits mention of agriculture as harboring a set of important social problems. It makes several crucial assumptions that are simply not true. In the first place, we know that any technology creates social problems, and it is not always the case that benefits outweigh costs.

In the second place, burgeoning populations will not assure consumption of any and all increases in food crop production. The truth is that, even though the world's population may double in 35 years, a large proportion of the populations of LDCs are and will continue to be unable to consume their share of increased food production because they lack purchasing power. Unemployment and underemployment rates are astronomical, and hence there are substantial numbers of people who are not sharing in any meaningful way in the economic growth of the past decades in LDCs. The recent picture of India exporting wheat while sizable numbers go hungry is simply an illustration of the point. The World Bank (1981:103) states: "the most common cause of undernutrition is a demand, not a supply, factor—a straightforward lack of purchasing power."

In the third place, to a large extent the poor are in small-scale agriculture, and in many countries must be relied upon to produce staples. If they are not granted needed factors of production, including appropriate technology and incentives, they may be unable to increase the supplies of farm commodities they produce adequately, and the system will increasingly tend to depend on imports.

In the fourth place, much of this small-scale agricultural work is done by women who have been consistently deprived of access to both basic education and the technical information they need—to say nothing of such other production factors as secure access to land, credit, etc.—to maintain current production levels.

Growth, Distribution, and Institutions in LDCs

So much of the success of "the U.S. model" in providing adequate food for the American consumer is based on the fact that our income distribution is not impossibly skewed (though it is more so than in many other DCs). Nearly everyone can buy enough food, and those who cannot have been helped to do so by consumer subsidy programs. Even the most recent U.S. administration has found it difficult to excise food stamps, school lunches, and other programs like them. As Lester Thurow (1981) concludes, until very recently "every group participated in advances after World War II so that there were no changes in the income shares going to different parts of the population. The poor neither gained nor lost relative to the rich."

On the other hand, when growth is concentrated in any enclave or "semi-enclave" in one society (as it is in many LDCs), and the trend to inequality is continuing, adoption of inappropriate technology is more possible, and the resultant costs to society are higher. Adoption may simply aggravate inequity problems. In this case, the key problem for development is not lack of growth—average figures have been rather remarkable in some LDCs lately. A key bottleneck is distribution, and it may be questioned whether growth itself can continue in the face of such inequality among and within nations (World Bank, 1978, 1981; Lewis, 1980).

From 1950 to 1975, the absolute disparity between the richest and the poorest developing nations increased threefold (Morawetz, 1977). Of course, gaps that separate DCs and LDCs became even more pronounced. More seriously for the current inquiry, within many less developed countries there is also a high degree of income inegalitarianism, with the upper 5 or 10 percent—which have the majority of the wealth, skill, and perhaps sheer influence—reaping the majority of the benefits, while the poorest, say, half of the population live at the margins of society and the bottom 20 percent loses ground.

It has been observed by many later writers, but Kuznets' early work led the way in pointing out that during industrialization income is bound to be concentrated for a time among the savers, the investors, the entrepreneurs, and the capitalists. But there seem to be two fundamental differences between the history of the already industrialized nations and contemporary LDCs: (1) During the industrialization of contemporary DCs there was a substantial group of individuals who could be counted upon to invest so that future generations could be employed; short-term concentrations of income resulted in more jobs in the medium or even longer term. The contemporary phenomenon seems not to be bringing such felicitous results. (2) Income in LDCs has been badly distributed for some time—for much longer than in contemporary DCs when they were industrializing—and there is little sign that surpluses are being opportunely invested: wealth is often held in the form of idle or underutilized land; unnecessary extravagances are the rule, not the exception; profits are removed from the country; plant and equipment are operating at under-capacity levels and with labor-saving technology. The poorest 50 percent or so of the LDC populations are characterized by large degrees of unemployment and underemployment. The LDCs suffer more than proportionally to conditions of worldwide recession which plague us in the early 1980s.

Alder (1972:331) puts the problem well: "the [growth] achievements of the last twenty years do not form a solid foundation for further and faster advances from now on—because of the uneven distribution of the benefits of economic growth, which has adversely affected the social and political cohesiveness of many countries and made them more prone to interruptions of economic progress through social upheaval and political turmoil."

Institutions reinforce and are reinforced by this inequality, sometimes ensuring that those farmers in the leading sector get what is "modern," while those in the lagging sector are poorly served by any technology. In this system, there is little or no "appropriate" technology.

The Case of Brazil

One of the fastest growing countries is worthy of a somewhat more detailed examination because it illustrates difficult problems of internal inequality.

Brazil's economy grew at an average annual rate of 8 percent between 1960 and 1970, and 10.6 percent between 1970 and 1976 (the Inter-American Development Bank places the 1977-80 average yearly growth rate at 6.3 percent). Most of this can be attributed to a rapidly growing manufacturing and services sector, but between 1970 and 1976 even agriculture grew by 5.5 percent (World Bank, 1978; IADB, 1982). Total population growth remained at a constant 2.9 percent throughout the period (World Bank, 1978; IADB, 1982). Even with this high rate of population growth, per capita income growth was significant indeed. Albert Fishlow, one of the foremost students of the Brazilian economy, states (1976:71): "This economic 'miracle' has already begun to rival the earlier German example. The common ingredient of greater scope to market forces and freer rein to the private sector has not gone unnoticed; there is already talk of the applicability of the Brazilian model to other parts of the developing world." Yet Fishlow argues that, on the basis of preliminary 1970 census data, the way in which the initial income increases and succeeding increments are distributed is a much more perplexing and troublesome matter. Using data not available when Fishlow wrote, but coming to about the same conclusions he reached years ago, Table 13.1 shows that while the poorest half of the workforce received 18 percent of total income in 1960, it received 15 percent in 1970, and this declined to 12 percent in 1976. The next 30 percent of the workforce fared no better; the share of the next 15 percent remained almost stable. Only the top 5 percent of the workforce made any real gains, moving from a 28 percent share of total income in 1960 to 35 percent in 1970, and 39 percent in 1976.

TABLE 13.1

Comparison of Distribution of Income in Brazil:
1960, 1970, and 1976

WORKFORCE	PARTICIPATION IN TOTAL INCOME		
(in percent)	1960	1970	1976
Poorest 50%	17.71	14.91	11.80
Next 30%	27.92	22.85	21.20
Next 15%	26.66	27.38	28.00
Highest 5%	27.69	34.86	39.00
Gini Coefficient	0.50	0.56	0.60

SOURCE: José Serra, "Renda concentra-se mais nos anos 70," Ensaios de Opiniao
(Rio de Janeiro, 1978), Tables 1, 2-6, p. 28.

In an examination of 1,000 farms in Brazil in the late 1960s, along with some
other Latin American data, Cline reinforces this view of the entire economy with
the observation that income distribution worsened in the decade. He concluded
that if specific policy measures were not devised to correct the situation, rural
income, and by implication all incomes, would continue to become more concentrated. Cline went further, suggesting some policy alternatives that had been or
were to be reiterated by others who examined the same or similar situations: (1)
land redistribution or agrarian reform is the policy most likely to bring both production gains and improved equity; (2) new crop varieties, highly desirable from
the standpoint of production alone, may have adverse income distribution
effects.[3] Cline also suggested that if big farmers adopted mechanization with
tractors or combines, the usual case would be that they would displace labor,
increase unemployment, and hence make worse the situation for those on the low
end of the spectrum of income receivers. He concluded that if output gains in
Brazil could be channeled to low income rural families, there would be little loss
of potential savings and production. More pointedly stated, Brazil could, with a
set of appropriate governmental policies, have more equity without much sacri-
fice in her rate of growth (Cline, 1973, 1970).

As we know, the specific policy measures that Cline called for were not
forthcoming in the rural sector of Brazil. Their analogues did not come in the
urban sector either, and consequently an enormous income gap between the
poorest 50 percent of the population and the top 5 percent appears by 1976, as
shown in Table 13.2. In constant *cruzeiros,* in 1960 the average earnings of the
top 5 percent group were 15 times higher than those of the bottom half; the
comparable coefficient in 1976 was about 33.

Of course, some have pointed out that even the poorest 50 percent have in
fact "participated" in the growth process because their income in 1976 was 66

TABLE 13.2

Median Income per Population Group in Brazil
(in constant cruzeiros)

WORKFORCE	PARTICIPATION IN TOTAL INCOME		
(in percent)	1960	1970	1976
Poorest 50%	73.4	84.8	140.4
Next 30%	192.7	216.7	420.0
Next 15%	396.1	519.3	1,109.7
Highest 5%	1,131.0	1,984.0	4,638.0

SOURCE: José Serra, "Renda concentra-se mais nos anos 70," Ensaios de Opiniao (Rio de Janeiro, 1978), p. 28.

percent greater than in 1960. By this reasoning, all have "participated" in that the "next 30%" had 94 percent more income in 1976; the "next 15%," 114 percent more; and the "highest 5%," 134 percent more.[4] The following arguments have been made in response to the idea that "everyone has benefited but some more than others": (1) as agriculture has modernized, more salary came to be paid to rural workers in cash rather than "in kind" (while they are a vestige of semi-feudalism, at least these "in kind" payments do not depreciate with inflation); (2) legislated minimum salaries in urban areas have risen to a greater extent than rural incomes; the fact that the plight of the rural worker is likely to be worse than that of his urban colleagues is masked by the aggregation of these data[5] (3) if the lowest 50 percent were more disaggregated, it would show that the bottom 20 percent have lost income in absolute terms, and these are probably largely rural people (Weisskoff and Figueroa, 1976).

Income Distribution in Selected LDCs: Cross-Sectional Data

The distribution of income in many LDCs with per capita incomes in the upper range is similar but less well documented (Brazil has exceptional facilities for gathering data). Income in the very poorest of nations is much more evenly distributed simply because any population needs a certain minimum income for survival. Cross-sectional income distribution data suggest a rough, asymmetrical "U" shape if "degrees of inequality" are plotted on a vertical axis and per capita income on a horizontal axis, with the poorest LDCs and the DCs at the more egalitarian ends of the curve. The egalitarian distribution of poverty in the poorest LDCs is disconcerting by any man's measure of humanitarianism. It is not much comfort to find countries where nearly everyone is equally destitute.

This "U" was shown clearly in cross-sectional data gathered by Adelman and Morris.[6] The point is illustrated further by the later United Nations data in Table 13.3, which groups countries into quintiles according to income classifications. In countries with per capita incomes either below $200 or above $2,000, income is more evenly distributed than in those countries between these two extremes.

TABLE 13.3

**Distribution of Income for Non-Socialist Countries,
Grouped by Level of GDP per Capita, in Quintiles***

GDP (in dollars)	NO. OF COUNTRIES	QUINTILES					HIGHEST 5% OF INCOME RECIPIENTS (incl. in Q_5)
		Q_1	Q_2	Q_3	Q_4	Q_5	
		(Percent)					
Below 100	9	7.0	10.0	13.1	19.4	50.5	29.1
101-200	8	5.3	8.6	12.0	17.5	56.5	24.9
201-300	11	4.8	8.0	11.3	18.1	57.7	32.0
301-500	9	4.5	7.9	12.3	18.0	57.4	30.0
501-1,000	6	5.1	8.9	13.9	22.1	50.1	25.4
1,001-2,000	10	4.7	10.5	15.9	22.2	46.6	20.9
2,001 and above	3	5.0	10.9	17.3	24.1	42.7	16.4

SOURCE: United Nations, Dept. of Economic and Social Affairs, *Report on the World Social Situation,* E/CN.5/512/Rev.1, ST/ESA/24 (New York, 1975), p. 14, based on Felix Paukert, *ILO Review* 108:2/3 (1973), p. 118.

*Coverage for individual countries ranges from 1948 to 1971.

Adoption and Use of Technology When Incomes are Unequal

It should stand to reason that if an economy has a lot of unskilled labor, some natural resources, and very little capital – characteristic of most LDCs – it will use a labor-intensive technology. But if the economy is divided and bipolar, such that one sector has these characteristics and the other concentrates the bulk of the skills and scarce capital, it does not necessarily follow that capital-intensive technology, readily available from DCs which are indeed anxious to sell it, will be kept out. (In large LDCs this technology is readily manufactured at home.) That this adoption happens in fact – especially in those middle-income LDCs where inequality is especially pronounced – flows directly from a situation of maldistribution. It is not enough to think of the capital-skills sector as a leading edge — the pioneering frontier — of economic development for there are political notions of injustice and power which make it a "privileged sector," one favored far beyond normal market expectations. Those in this elite position are, for example, able to get subsidies for the importation of capital-intensive technology which most suits them in the form of concessionary credit and can take the most favorable advantage of differential foreign exchange rates. Sometimes even corruption and bribery come to be important to a greater extent than when more equality, and the accountability that often accompanies it, is more the rule. Adoption costs for inappropriate technology rest not solely with those who ultimately utilize the technology but on the society as a whole. All of society, and disproportionately the poor, must pay the dislocation costs of widespread unemployment and the higher prices in manufacturing for a lower quality of consumer goods (protection is also usually part of the package) that can compete only in the sheltered home market.

Under conditions of more equality, if the economy has a "leading sector," it is connected through the factor market to the "lagging sector," and this symbiotic relationship (the leading sector needs labor; the lagging sector needs wages) creates at least some "trickle down" to the backward sector. There is a forward connection to the output market also, for wage workers are also consumers.

If in LDCs capital-intensive technology is used in the leading sector, the factor market connection between the two sectors becomes constricted, the sectors become progressively less integrated, and resultant income and wealth become even more concentrated.

As less labor is used, there will be an impact on the size of the market. A smaller wage bill means little income to serve as demand for purchased goods to soak up the excess capacity from agriculture or manufacturing, and this in turn increases protectionism even more and adds further to the economic isolation of the lagging sector.

Hans Singer notes that, if utilizing capital-intensive technology means preservation of a bipolar society (1977:9), "the problem can be seen to be one of *equality*—i.e., whether access to scarce capital and skills should be fairly evenly spread over the whole economy or whether it should be concentrated in one small limited sector to the exclusion of the rest. . ."

When technology is "appropriate" in LDCs it is often also highly divisible (see Dorner's chapter in this volume). A good example is the seeds and accompanying fertilizer of the Green Revolution. When Azam analyzed the Pakistan Punjab (1973), he concluded that, while the smaller farms do face relatively more severe constraints of not-so-divisible irrigation water and credit, "the differences in the severity of these constraints is not serious enough to have caused any significant differences in yields obtained by the small farmers as compared with the large farmers." Similar results have been reported, according to Ruttan (1977), for the Philippines and Indonesia. Ruttan reports further that the introduction of new high-yielding wheat and rice technology has resulted in a second-round increase in the demand for labor, a finding also of Johnston and Cownie (1969).

The points made here are that when the economic structure of the country is bipolar, inappropriate technology may be adopted by the privileged sector, and this technology exacerbates the income, skills, and capital gaps between the two sectors. Appropriate technology may help to break down this sharp dualism. In analyzing technology's distribution through institutions, one must also be aware of how technology may change those institutions. The feedback loops from technology adoption may be important determinants of institutional change, an idea that is Hayami and Ruttan's (1971) contribution to this thinking.

While agriculture's privilege in LDCs can take advantage of monopolistic elements and political power, some Green Revolution technology has been divisible enough so that it does have some impact on the lagging segment of agriculture. But all producers of agricultural products usually have to sell in a perfectly competitive market for the same reason that U.S. producers do. And this is one reason for Ruttan's conclusion (1977:19) that the effect of high yield varieties has been to dampen the rate of increase of food grain prices, thus giving enormous benefit to domestic consumers, especially poor ones:

> In Asia over 60 per cent of the wheat area and over one quarter of the rice area are planted to the modern varieties developed since the mid-1960s. . . [I]n crop year 1974-75, the supply of rice in all developing countries was approximately 12 per cent higher than it would have been if the same total resources had been devoted to production of rice using only the traditional rice varieties available prior to the mid-1960s.
>
> The impact of a shift to the right in the supply of food grain is particularly significant for both the urban and rural poor. The distribution of grains among consumers depends primarily on the relative amount of a particular commodity consumed by each income stratum and on the price elasticity of demand in each stratum. The larger the quantity consumed and the higher the absolute value of the price elasticity of demand in the lower income strata, relative to the higher income strata, the more favorable will be the distributional benefits.
>
> This is illustrated quite dramatically by the impact of the new rice technology on consumer welfare in Colombia since the mid-1960s. Between 1966 and 1974 the percentage of the area planted to modern varieties rose from 10 per cent to 99 per cent. Yields on irrigated land rose from 3.1 to 5.4 metric tons per hectare, and total rice production increased from 600 thousand to 1,570 thousand metric tons. Most of the increased production was absorbed in the local market. The benefits were transmitted to consumers through both lower prices and increased per capita consumption.
>
> The benefits were strongly biased in favor of low income consumers. The lowest income quartile of Colombian households, which received only 4 per cent of household income, captured 28 per cent of the consumer benefits resulting from the shift to the right in the supply curve for rice.

Why Adopt Inappropriate Technology?

Adoption or encouragement of technology, therefore, can be seen to have profound political implications. Inappropriate technology is not adopted blindly, irrationally, or by accident, but because it makes economic sense to at least one class or sector and the remainder of the society acquiesces for lack of power. If only an enclave or privileged sector adopts, it also serves to make that group more of a freestanding enclave and less of an integrated part of an interdependent economy. In time, we might surmise, it serves to make that group more vulnerable politically, as, for example, the upper classes in countries such as El Salvador and Guatemala in the late 1970s and early 1980s. Likewise, if appropriate technology were to be fostered in LDCs, it could strengthen underclasses (Weiss, Jr., 1979:1083).

Some blame the DCs for foisting their technology on unwilling LDCs. Singer (1977) suggests, "what is wrong is not that we in industrial countries have developed an efficient technology of the kind right for us—who could blame us for this? But what *is* wrong is that there is not an equally efficient technology of a *different* kind helping the Third World countries to become rich, just as we in our time were helped to become rich by a technology quite different from the one which suits our purposes now. Thus the primary need is to develop technological power and capacity in the developing countries."

The development of an appropriate technology does not mean a return to a "second best," nor is it a less intellectually demanding task. In developing appropriate technology lies a new frontier of knowledge, a genuine forward movement into new territory. But any discussion of this type would be remiss not to recognize that:

> As a practical matter, governments of developing countries usually control the introduction of innovations . . . . However, these governments are, with a few exceptions, not monoliths. For this reason, there will be times when a successful technological experiment can be the vanguard of social change. Development of small-scale energy sources may open up new possibilities for rural development in areas remote from power grids. Development of technologies for small-scale farming or industry may undermine the justification for exclusive reliance on large-scale technology. Research or demonstration projects may thus highlight the existence of alternative strategies as a first step toward mobilizing public opinion or readying official opinion for change. (Weiss, Jr., 1979:1087)

As an example of small-scale energy sources, Vietmeyer (1981) recently noted that animal power is a type of appropriate technology in many countries but is almost entirely overlooked as a fit subject for research:

> Even now, after we've harnessed steam, oil, and electricity, animals contribute about half the energy used for agriculture in the third world. Four-legged traction

provides some developing countries with as much as 90 percent of their agricultural power. . .

"India has put a satellite in space and harnessed the atom," says N.S. Ramaswamy, director of the Indian Institute of Management, in Bangalore, "but our carts are 5,000 years old, because professors are scared they may not get promoted if they work on designing better ones. . . ."

In the 1980s, we could make. . .advance[s] by improving the yokes, harnesses, implements, and vehicles used with the millions of oxen, water buffaloes, and other animals. A relatively small effort would be enough to dramatically improve their efficiency. For the people below the poverty line, it would be the biggest economic improvement we could make. All that is holding it up is the closed mindedness of educated people.

The pressing issue of LDCs today is that income is too concentrated in too few hands while the vast majority are landless and jobless. The question is deceptively simple: Can this situation be reversed before it becomes irremediable? Developing appropriate technology may well be at least part of the answer.

Notes

1. Peter Dorner states, "U.S. agriculture was transformed from a largely self-sufficient system to one highly productive and deeply interdependent with the rest of the economy. In this respect, the United States has travelled the road over which all nations must pass as they seek accelerated growth and broad-based economic development." But, he concludes, "a comparable course of agricultural development would not likely to be merely problematic for most countries where capital is scarce and labor is in over-abundant supply, *it would be disastrous!*" Dorner's paper for the World Conference on Agrarian Reform and Rural Development in Rome, July 1979, is a concise and illuminating history of the development of U.S. agriculture. "Rural Development Problems and Policies: The United States Experience," mimeographed, Rome, February 1979, pp. 2 and 47. His analysis of the systems' weaknesses (pp. 40-44) comprise an effective summary.
2. Of course, some farms in the United States are not "family farms" and never were; others have recently become "corporative." Some of this can be traced to the historic settlement pattern and subsequent development pattern of California, Arizona, and other sections of the Southwest and West which were quite different from that of the Midwest and East. Likewise, parts of the Gulf Coast and Florida have a rather unique heritage. Also, we must not forget that "pockets of poverty" exist as some agricultural areas have been left behind by inexorable technological change.
3. He seems to have been assuming that big farmers would adopt first and small ones perhaps not at all, so that as prices fell, small farmers would be squeezed out of business into a landless category.
4. Gary Fields, "Who Benefits from Development? – A Reexamination of Brazilian Growth in the 1960s," *American Economic Review* 67(September, 1977):570-581, disagrees with the Serra interpretation partially on these grounds.

5. Personal conversation with Fernando Dell'Acqua; and other articles in *Ensaios de Opiniao,* previously cited.
6. I. Adelman and C.T. Morris, *Economic Growth and Social Equity in Developing Countries* (Stanford, CA: Stanford University Press, 1973); it is traced to original works by Kuznets and most recently has appeared again in Montek S. Ahluwalia, "Inequality, Poverty, and Development," *Journal of Development Economics* 3 (1976): 307-341.

References

Adler, John H. 1972. "Development and redistribution." World Bank Reprints no. 8, from *Weltwirtschaftliches Archiv* 108:331.

Azam, K.G. 1973. "The future of the green revolution in West Pakistan: a choice of strategy." *International Journal of Agrarian Affairs* 5(March): 404-429.

Chambers, Robert. 1977. "Appropriate to what?" *Ceres* 10(May-June):27-30.

Cline, William R. 1973. "Interrelationships between agricultural strategy and rural income distribution. *Food Research Institute Studies in Agricultural Economics, Trade, and Development* 12(2):138-157

_____. 1970. *Economic Consequences of a Land Reform in Brazil.* Amsterdam: North Holland.

Fishlow, Albert. 1976. "Brazilian size distribution of income." In Alejandro Foxley (ed.), *Income Distribution in Latin America.* Cambridge, England: Cambridge University Press.

Hayami, Yujiro, and Vernon W. Ruttan. 1971. *Agricultural Development: An International Perspective.* Baltimore: Johns Hopkins University Press.

Howes, Michael. 1979. "Appropriate technology: a critical evaluation of the concept and the movement." *Development and Change* 10(January):115-124.

Hulse, Joseph H. 1982. "Food science and nutrition: the gulf between rich and poor." *Science* 216(18 June):1291-1294.

Inter-American Development Bank (IADB). 1982. *Economic and Social Progress in Latin America, 1980-81.*

Jedlicka, Allen D. 1977. *Organization for Rural Development: Risk Taking and Appropriate Technology.* New York: Praeger.

Johnston, Bruce F., and J. Cownie. 1969. "The seed-fertilizer revolution and labor absorption." *American Economic Review* 59(September):569-582.

Lewis, W. Arthur. 1980. "The slowing down of the engine of growth." *American Economic Review* 70(September):555-564.

Morawetz, David. 1977. *Twenty-Five Years of Economic Development: 1950 to 1975.* Baltimore: Johns Hopkins University Press, for World Bank.

Press, Frank. 1978. "Science and technology: the road ahead." *Science* 200(19 May): 737-741.

Ruttan, Vernon W. 1977. "The green revolution: seven generalizations." *International Development Review* 19(4):16-23.

Schumacher, E.F. 1971. "Industrialization through 'intermediate' technology." Pp. 85-94 in Ronald Robinson (ed.), *Developing the Third World: The Experience of the Nineteen-Sixties.* Cambridge, England: Cambridge University Press.

Singer, Hans W. 1977. "Appropriate technology for a basic human needs strategy." *International Development Review* 19(2):8-11.

Swanberg, Kenneth G. 1980. "Small farmer technology adaptation: reducing the constraints caused by the requirements-limitations gap." *Development Discussion Paper Series* no. 88. Cambridge, MA: Harvard Institute for International Development.

Thurow, Lester. 1981. "The widening income gap." *Newsweek* (5 October).

Vietmeyer, Noel. 1981. "Animal power." *New York Times* (10 December).

Weiss, Jr., Charles. 1979. "Mobilizing technology for developing countries." *Science* 203(16 March):1083-1089.

Weisskoff, R., and A. Figueroa. 1976. "Traversing the social pyramid." *Latin American Research Review* 11(2):71-112.

World Bank. 1981. *World Development Report.* Washington.

————. 1978. *World Development Report.* Washington.

Index

Contributors

Frederick H. Buttel
    Department of Rural Sociology
    Cornell University
    Ithaca, New York

Peter Dorner
    Dean of International Studies and Programs
    University of Wisconsin
    Madison, Wisconsin

Frederick C. Fliegel
    Department of Agricultural Economics
    University of Illinois
    Urbana, Illinois

Boguslaw Galeski
    Institute of Philosophy and Sociology
    Polish Academy of Science
    Warsaw, Poland

John W. Gartrell
    Department of Sociology
    University of Alberta
    Edmonton, Canada

Wava Gillespie Haney
    Land Tenure Center
    University of Wisconsin
    Madison, Wisconsin

Sheldon Krimsky
    Department of Urban and Environmental Policy
    Tufts University
    Medford, Massachusetts

Allan Mazur
    Department of Sociology
    Syracuse University
    Syracuse, New York

David A. McGranahan
    Economic Research Service
    U.S. Department of Agriculture
    Washington, D.C.

Denton E. Morrison
    Department of Sociology
    Michigan State University
    East Lansing, Michigan

Howard Newby
    Dean, Faculty of Arts and Letters
    University of Essex
    Colchester, England
        and
    Department of Rural Sociology
    University of Wisconsin
    Madison, Wisconsin

Allan Schnaiberg
    Department of Sociology
    Northwestern University
    Evanston, Illinois

William C. Thiesenhusen
    Director, Land Tenure Center
    University of Wisconsin
    Madison, Wisconsin

Johannes C. van Es
    Department of Agricultural Economics
    University of Illinois
    Urbana, Illinois